BRITISH RED DATA BOOKS : 1

VASCULAR PLANTS

Compiled by

F. H. PERRING and L. FARRELL

Biological Records Centre
Institute of Terrestrial Ecology

(Badger)　　　　(Frog orchid)　　　　(Panda)

Published by

THE SOCIETY FOR THE PROMOTION OF NATURE CONSERVATION

with the financial support of THE WORLD WILDLIFE FUND

1977

SPNC
The Green
Nettleham
Lincoln

Library of Congress Catalog Card Number:

ISBN 0 902484 02 8

ISSN 0140-1122

The SPNC wishes to acknowledge the generous help given in preparing this book by the Biological Records Centre of the Institute of Terrestrial Ecology, Monks Wood Experimental Station, Huntingdon, under contract to the Nature Conservancy Council and the financial assistance of the World Wildlife Fund without which this publication would not have been possible.

CONTENTS

FOREWORD

There are at least two components of effective and discriminating policies for conserving plant species and plant assemblages. First, it is essential to know the extent of the resource, and, second, it is desirable to have an understanding of how the resource will react to different management prescriptions aimed at sustaining it.

Using the extensive records collated by staff of the Biological Records Centre at the Institute of Terrestrial Ecology's Monks Wood Research Station, it has been possible to construct a list of rare or threatened species now occurring in 15 or fewer 10 kilometre squares of Great Britain. On checking, some species thought to be rare are not included in this category. Nevertheless there are 321 species or subspecies in the rare or threatened category, including an increasing number of species which were formerly weeds of arable areas.

The Biological Records Centre, with the help of the many recorders of the Botanical Society of the British Isles, has succeeded in establishing the extent to which species should be regarded as rare or threatened. We now need to learn more about the biology and ecology of these species, including their responses to competing plants. With this knowledge, it should be possible to forsake the temporary and limited expedient of preservation, and establish an active, effective and integrated policy of conservation.

J. N. R. Jeffers
Director,
Institute of Terrestrial Ecology.
2nd June 1977

i

INTRODUCTION

In the field of nature conservation it is becoming increasingly important to be able to balance legitimate developments of industry and agriculture on the one hand with the conservation of the environment on the other: wise decisions demand correct facts.

Whilst nature conservation in Britain grew out of concern for the future of ecosystems characteristic of our country there has been a steady increase in recent years in concern for the survival of species per se. For species endemic to these islands we have a clear responsibility to the scientific world to ensure that none is lost. But many of our non-endemic species are represented by populations which have been isolated from continental populations for several thousand years, and exhibit morphological differences which have been recognised at subspecific or varietal rank: this variation must also be saved.

Ideally the conservation of species would be a by-product of the conservation of ecosystems and whilst it is true that the policy of acquiring nature reserves in Britain in relation to their contribution to ecosystem conservation has resulted in many of our rare plant species now receiving adequate protection — this has not been successful in protecting all the sites of all our rare and endemic species. A recent survey of twenty of the rarer species likely to be protected by legislation showed that out of the 231 sites in which they occurred only 85 were in nature reserves.

Thus for an efficient nature conservation policy guidance is required to the species most in need of protection so that priorities for the acquisition of new reserves and for other conservation measures can be objectively established. Whilst for classically rare species there are no problems of selecting which to include once rarity has been defined, it is known that many once widespread species have declined dramatically since recording began and it is possible that the decline continues. These species too are candidates for an active conservation policy but without recent and almost continuous survey the list of which should be included would be extremely arbitrary.

This book is an attempt to make a factual statement about the present status of the endemic and rare vascular plants in Britain and to include those other species which appear to be now so rare that their continued existence in our flora is in question. By no means all the species included are endangered but we feel that an objective document should provide evidence capable of assisting conservationists not only to take action where it is urgently required but to resist pressures to act when no action is really necessary.

If this work becomes recognised as the equivalent of an International Union for the Conservation of Nature and Natural Resources Red Data Book for British Plants then it should be appreciated that if red is for danger then not all the species included are threatened and that for some the colour is such a pale pink as to be scarcely separable from white. We believe though that with the exception of some critical genera (see p. vi) all species of British flowering plants and ferns which are truly endangered have been included.

BACKGROUND OF SURVEY

If the 1950s was the decade in which the foundation of nature conservation was laid in Britain then the 1960s was the decade in which conservation went onto the offensive — converting the people of Britain from a nation which took wildlife for granted to one which was deeply concerned about its future. This concern was crystallised in the enormous impact of European Conservation Year 1970. In this context it was apparent that conservationists had a duty to examine the biological richness of the nation and state firmly what should be conserved at all costs.

In 1968 the *Critical Supplement of the Atlas of the British Flora* was published: the final result of the Botanical Society of the British Isles Distribution Maps Scheme begun in 1954, which had resulted in the *Atlas of the British Flora* (1962) six years earlier. By this time the Maps Scheme had become the foundation of the Biological Records Centre of the then Nature Conservancy's Experimental Station at Monks Wood near Huntingdon and the publication of the *Critical Supplement* released the botanical staff, at that time Miss M. N. Hamilton and F. H. Perring, to consider the most appropriate direction our future work on the British flora should take.

In the context of the time we had no doubt that we should use our data and the network of recorders on which it was based to collect as much information as possible about the present distribution of the rarer species in the flora and to compare this with the known distribution of those species in the past so as to present an accurate account of changes taking place in that segment of our flora.

This choice of study became even more relevant because of an attempt made by a private member, Mr. Peter Mills, M.P., in January 1968 to get a Wild Plants Protection Bill through the Houses of Parliament. There was an immediate demand by the Nature Conservancy for information about the species which were included in the first and second schedules of that Bill and a report was prepared for them.

This attempt failed but a summary of the report was presented in a paper for the 1969 BSBI Conference on the *Flora of a Changing Britain* and published a year later (Perring, 1970). Whilst the figures were incomplete, being based mainly on the work done during the preparation of the *Atlas* (which accepted as recent, records for some areas made in the 1930s) they indicated that as many as two thirds of the localities of our 300 rarest species had probably been lost since recording began in Britain over 300 years ago, and that the rate of extinction was increasing.

The presentation of these interim results coincided with the publication in Belgium of *Plantes rares, disparues ou menacées de disparition en Belgique* (Delvosalle *et al*, 1969). Not only did this detailed work show that the losses in Belgium were as great or greater than those in Britain, but it served as an inspiration for the kind of work which could be invaluable to conservation organisations in this country.

At this point it became imperative that we produce a Red Data Book for flowering plants and ferns with all possible haste.

PRODUCTION OF THE RED DATA BOOK

AREA OF SURVEY

On biogeographical grounds it is difficult to justify the preparation of a single list of species for the British Isles as a whole. Many of the rarest species in Ireland, where the conservation of their isolated populations may be essential, are very abundant in Britain, and there is also a considerable element in the flora of Ireland which does not occur in the rest of Britain, so a separate list of Irish species seems desirable and is being prepared by the Irish Committee of the BSBI and the Irish Biological Records Centre. The flora of the Channel Isles is closely related to that of the adjacent coast of France and it would not be sensible to publish, in a list of rare British plants, the considerable number which occur in the British Isles only in the Channel Isles, but are nearly all abundant in north-west France.

The Nature Conservancy Council (formerly the Nature Conservancy) extends its executive powers to Great Britain: the SPNC (the Association of Nature Conservation Trusts) co-ordinates the activities of the 40 Nature Conservation Trusts in Great Britain. Neither organisation extends to Ireland or the Channel Isles.

So for scientific and administrative reasons this list is limited to species which are rare in Great Britain. For completeness, however, a statement is included of the present status of these species in the Channel Isles and Ireland as far as it is known.

THE SPECIES INCLUDED

The initial selection of species was made by inspection of the *Atlas of the British Flora* and the *Critical Supplement*: all native or probably native species which were recorded in 15 or fewer 10 kilometre squares from 1930 onwards were included. However most of the large critical genera have been omitted notably *Hieracium* and *Rubus* though these may be the subject of a subsequent volume.

During the course of the investigations, which are described below, 42 taxa which were found to occur now in more than 15 10 kilometre squares were discarded and these are listed in Appendix I.

At the same time it became apparent that some species, particularly arable weeds, which had occurred in over 15 10 kilometre squares when the *Atlas* was being prepared have declined so rapidly that they might now qualify for inclusion. For this reason, early in 1976 a request was sent out to BSBI vice-county recorders to send in all records known to them since 1960 of the 38 species listed below:

Agrostemma githago   *Limosella aquatica*
Ajuga chamaepitys   *Melittis melissophyllum*
*Alopecurus bulbosus   **Mentha pulegium*
*Bupleurum rotundifolium   *Myriophyllum verticillatum*
*Campanula patula   **Oenothera stricta*
*C. rapunculus   *Orchis ustulata*
Caucalis latifolia   *Polygonum dumetorum*
*C. platycarpos   *P. mite*
*Centaurea calcitrapa   *Ranunculus parviflorus*
C. solstitialis   *Rosa agrestis*
*Chenopodium vulvaria   *R. elliptica*
Cuscuta europaea   *Scandix pecten-veneris*
Dianthus armeria   **Sisymbrium irio*
Euphorbia platyphyllos   *Torilis arvensis*
*Filago lutescens   *Turritis glabra*
*F. pyramidata   *Valerianella dentata*
*Galium tricornutum   **V. rimosa*
*Gastridium ventricosum   *Viola lactea*
*Lactuca saligna   *Wolffia arrhiza*

Within the next twelve months returns were received from 90 per cent of the vice-counties in England, which is the part of Britain in which the arable weeds mainly occurred. These returns together with records already at BRC or abstracted from recent County Floras indicated that the 17 species marked with an asterisk in the list should be included because it is almost certain that all of them now occur in 15 or fewer 10 kilometre squares. *Caucalis latifolia* was omitted because it is apparent it was never more than a casual: *Rosa agrestis* and *R. elliptica* were omitted because of insufficient information.

To the list are added five species which have only recently been recognised as occurring in Great Britain but which apparently have only small populations:

Atriplex longipes   *Galium fleurotii*
A. praecox   *Ophrys bertolonii*
Gagea saxatilis

The final list published here consists of 321 species or subspecies which represents about 18 per cent of our native or probably native flora.

THE CONDUCT OF THE SURVEY

In 1968 the Biological Records Centre sent each vice-county Recorder of the BSBI a list of rare species for their area arranged in 10 kilometre square order, with a request that they inform the BRC of the present status of each plant in each locality listed, and add any new localities known to them which had been omitted. Up-to-date records were particularly required.

The response from the BSBI recorders was as good as could be expected but in the time available was inevitably dependent on their existing knowledge and could not be based on a new survey in the field.

Consequently, though the returns were invaluable in providing an estimate of the extent of the changes in the distribution of our rare species which had taken place, it was decided that this survey should be followed by one in greater detail in which an attempt would be made to collect precise and up-to-date information not only on the location of the species, but on the size and exact boundary of each population, upon the biology of the species at the site, and upon the conservation status of the area. A new 'Population Form' was designed (Fig. 1) and once again the BSBI recorders were asked to collaborate, and to their great credit many of them responded most willingly and have provided the majority of the 1,100 forms completed during the period 1970-76.

However in addition many individual members of the BSBI have given information about particular populations or have accepted our invitation to visit sites for which no population forms had been completed, whilst the Society itself ran several field meetings specifically directed to rare species. In 1972 the World Wildlife Fund, in collaboration with Pedigree Petfoods initiated a fund specially devoted to the conservation of rare or endangered British plants and animals. To this fund several applications were successfully made to support survey work on rare plants, and as a result botanically rich but remote areas in north Scotland, the Pennines and south-west England received extra attention. The Fund also made possible several autecological studies of endangered plants including *Lathyrus palustris* and *Viola percisifolia*.

A further, and most significant contribution was made in 1974/75 as a result of a contract made by the Nature Conservancy Council to the University Botanic Garden, Cambridge, which has provided for the salaries of a scientific officer and a propagator who have between them collected population data for rare species in eastern England and set up a bank of living material of endangered taxa in the Botanic Garden.

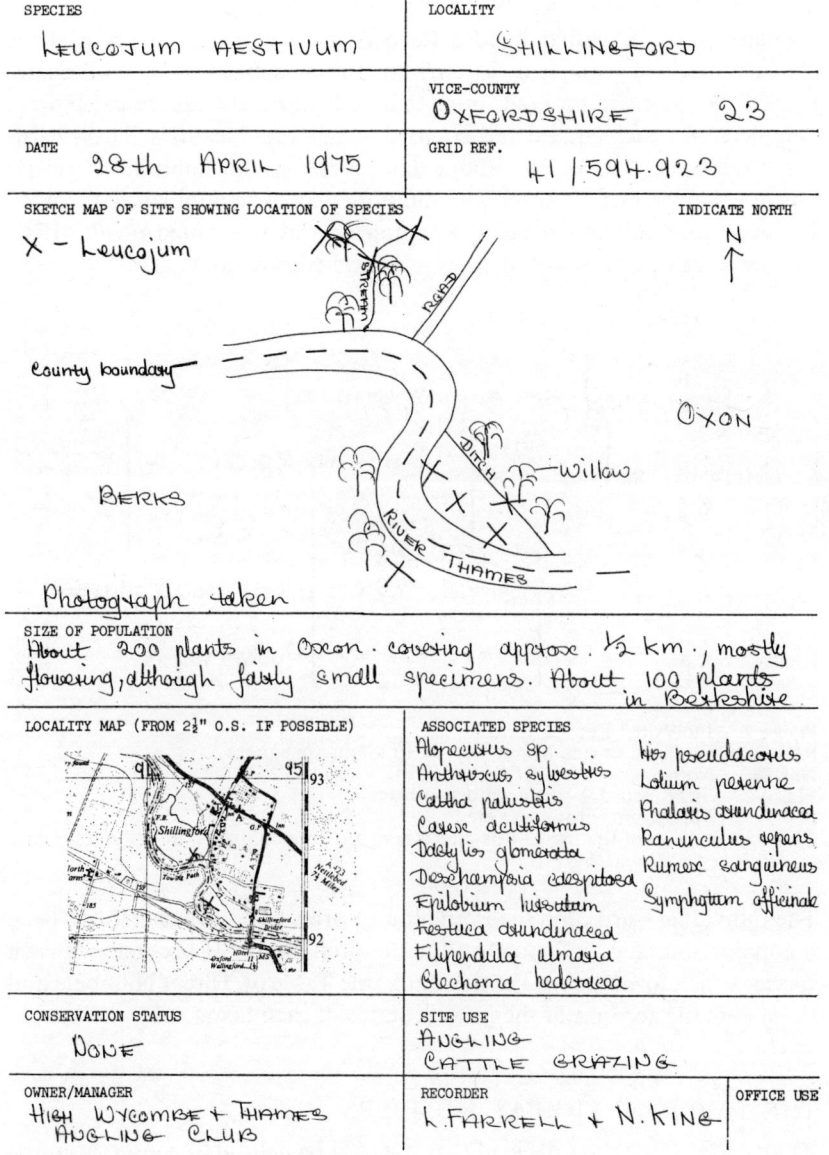

SPECIES	LOCALITY
LEUCOJUM AESTIVUM	SHILLINGFORD

VICE-COUNTY
OXFORDSHIRE 23

DATE 28th APRIL 1975

GRID REF. 41/594.923

SKETCH MAP OF SITE SHOWING LOCATION OF SPECIES   **INDICATE NORTH**

X – Leucojum

county boundary

BERKS

OXON

Willow

Photograph taken

SIZE OF POPULATION
About 200 plants in Oxon. covering approx. ½ km., mostly flowering, although fairly small specimens. About 100 plants in Berkshire.

LOCALITY MAP (FROM 2½" O.S. IF POSSIBLE)

ASSOCIATED SPECIES
Alopecurus sp.
Anthriscus sylvestris
Caltha palustris
Carex acutiformis
Dactylis glomerata
Deschampsia caespitosa
Epilobium hirsutum
Festuca arundinacea
Filipendula ulmaria
Glechoma hederacea
Iris pseudacorus
Lolium perenne
Phalaris arundinacea
Ranunculus repens
Rumex sanguineus
Symphytum officinale

CONSERVATION STATUS
NONE

SITE USE
ANGLING
CATTLE GRAZING

OWNER/MANAGER
HIGH WYCOMBE + THAMES ANGLING CLUB

RECORDER
L. FARRELL + N. KING

OFFICE USE

Figure 1.
Population Form.

PROCESSING OF THE RECORDS

The record for every locality known to BRC has been transferred to a standard pink Individual Record Card (Fig. 2). For this survey and the statistics derived from it a 'locality' has been defined as a 1 kilometre square. When a site covered more than a 1 kilometre square cards were completed for each square and counted as separate localities. These cards carry the basic information about date, location and habitat, as well as any notes made by a reporter who did not complete a population form. If, however, a population form was completed this was noted on the IRC, but the extra details on the form were not transferred to it.

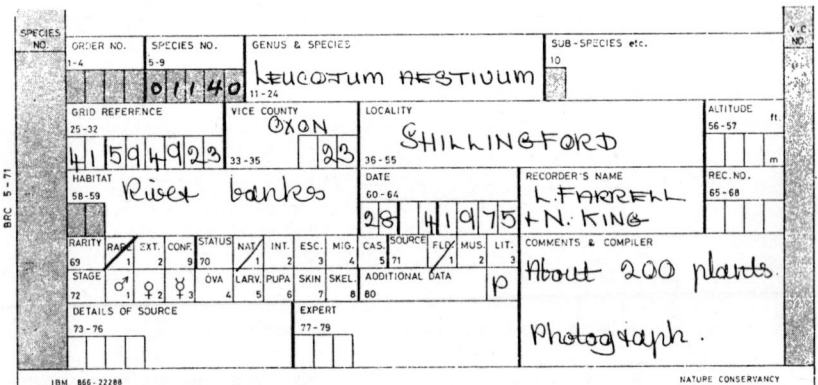

Figure 2. Individual Record Card
Rare = Confined to one locality in the 10 kilometre square
Nat. = Native
Fld. = Field record — no specimen collected
P = Population form completed
(Full explanations of the use of this and other cards are contained in B.R.C. booklet *Instructions for Recorders.*)

The Individual Cards and the Population Forms have, between them, been the major source of information for the preparation of accounts of each species which form the core of this book, the Table of Threat Numbers and the systematic account of the present status of each taxon.

THE TABLE OF THREAT NUMBERS

At an early stage it was decided that it would be helpful to conservationists if each species could be given a figure which would make it possible to arrange rare and endangered species in some order of priority. There have been many loose statements about the meaning of 'rare' and 'endangered' in the past and this is an attempt to be somewhat more objective. However the Threat Number here given still contains subjective elements (attractiveness, remoteness and accessibility) and the use of arbitrary categories (the Conservation Index). This is inevitable in a pioneer work of this kind. We hope that experience in the use of the Threat Number for this and other groups of plants and animals will produce refinements which will increase its value as a conservation tool in the future. Even then it is doubtful whether a single figure can ever be satisfactory in itself, and it is for this reason that the full information on how the Threat Number was arrived at is given in Table 1 which appears from pp. 82-92.

The Table is divided into nine columns:

Columns 1 and 2 show which species also occur in Ireland (H) and the Channel Isles (S) although the information in these two columns does not contribute to the Threat Number. The figures indicate present and past distribution in terms of 10 kilometre squares. Thus H 16/40 means that this species has been recorded from only 16 10 kilometre squares in Ireland recently, whereas it was recorded from 40 squares in the past.

In the Channel Isles five 10 kilometre squares are recognised, one for each of the five major islands, Alderney, Guernsey, Jersey, Herm and Sark. Thus under S, 3/4 means that this species has been recorded from four of the major islands in the past but from only three recently.

Column 3 refers to Great Britain (GB) and is divided into two sub-columns. To the left are figures which indicate present and past distribution in numbers of 10 kilometre squares. To the right is a figure which contributes to the Threat Number which is based on the data in the left hand column:

        0 = Decline of less than 33 per cent.
        1 = Decline from 33 to 66 per cent inclusive.
        2 = Decline over 66 per cent.

The number of 'recent' 10 kilometre squares given in columns 1-3 refers to a particular decade, details of which are given under the species account.

Ideally the same decade would have been used for each species: if all the sites of all the species had been visited since 1970 this decade could have been used throughout. In practice, though, whilst the rarer and more accessible species are visited frequently, the more widespread and inaccessible species may be recorded only very occasionally. Thus to have received no post 1970 record of a plant confined to Surrey would be extremely significant and cause for concern, but to have no records of a plant from a remote part of the Scottish highlands during ths same period would not necessarily be cause for alarm. The decade has been chosen in an endeavour to reflect real change rather than lack of observers.

A second edition of the *Atlas of the British Flora* which shows distribution in terms of 10 kilometre squares was published in 1976. In its preparation particular attention was paid to updating the maps of the rare and endangered species listed in this book: in general agreement will be found between the number of recent 10 kilometre square records in both books but this will not always be the case. In the *Atlas* the date line for recent records was usually 1930: in the *Red Data Book* the most recent decade which is practical has been used, as explained above. This is the usual reason for differences between the number of squares in these works.

Column 4 refers to Great Britain and is also divided into two sub-columns. The column to the left gives the number of extant localities of the species known to BRC. A locality being defined as explained on p.x above: in effect the number of 1 kilometre squares in which it has been recorded.

To the right is a figure which contributes to the Threat Number and has been derived from the data in the left hand column:

$$0 = \text{16 or more localities.}$$
$$1 = \text{10-15 localities.}$$
$$2 = \text{6-9 localities.}$$
$$3 = \text{3-5 localities.}$$
$$4 = \text{1-2 localities.}$$

Column 5 is a subjective assessment of the attractiveness of the species and the likelihood that it will be exposed to collecting pressure, particularly for the vase or garden. There are three categories:

$$0 = \text{not attractive.}$$
$$1 = \text{moderately attractive.}$$
$$2 = \text{highly attractive.}$$

In this context an insignificant plant like *Isoetes histrix* is in category 0, whereas a showy species such as *Potentilla fruticosa* falls in category 2.

Column 6, the Conservation Index, is an arbitrary figure related to the percentage of the localities of the species which are within nature reserves. An attempt was made throughout the survey to determine the conservation status of each site and on this basis the species is placed in one of four categories:

$0 =$ More than 66 per cent of localities in nature reserves.

$1 =$ From 33 to 66 per cent of localities in nature reserves.

$2 =$ Less than 33 per cent of localities in nature reserves.

$3 =$ Less than 33 per cent of localities in nature reserves and where these sites are subject to exceptional threat.

Columns 7 and 8 give subjective weighting to the relative ease with which the species can be reached by the public. The figure in *Column 7* (remoteness) results from considering whether the localities of the species as a whole are easy to reach by the public in terms of distance, comparing, for example, the threat between a species on a remote Scottish island and one on the North Downs within 20 miles of London.

The figure in *Column 8* (accessibility) results from considering the ease with which the localities of the species can be reached by the public when they have arrived at the site. Thus a plant of inaccessible cliffs is less threatened than one which grows by the roadside.

In both columns the following scale is used:

> 0 = not easily reached.
> 1 = moderately easily reached.
> 2 = easily reached.

Column 9 is the Threat Number and is the result of adding Columns 3-8. The maximum value is 15. No species has achieved this figure: the range arrived at is from 2-13.

A list of all the species arranged in Threat Number order and including extinct species appears as Table 1.

Column 10 is the IUCN Red Data Book Category (see p. xvii).

> EX = Extinct, E = Endangered, V = Vulnerable, R = Rare

THE SPECIES ACCOUNTS

The text is intended to convey the main areas in which the species occurs and any other areas where it grew in the past: to indicate the size of the populations of the species where this is known, and to give any evidence available on the causes of decline if this has occurred. In describing distribution the old pre-1974 county names have been retained. The new names, particularly in Scotland and Wales, cover such large areas that their use would have made the information given unnecessarily imprecise: it is also almost certain that these new boundaries will be altered yet again. Many biologists might have preferred the vice-county system but many of the vice-county names are unfamiliar and maps showing their boundaries are not generally available. A map showing the pre-1974 county boundaries is included (Fig. 3).

Each account includes at the bottom the past and present distribution in numbers of 10 kilometre squares from the Threat Number Table, and the Threat Number (TN) itself. A summary of the numbers of nature reserves and other sites of conservation status on which it occurs is also given.

We believe that the detail given in these accounts will in general prove adequate for discussing conservation problems with owners, planners and others involved in threats to the species. It was of course essential not to

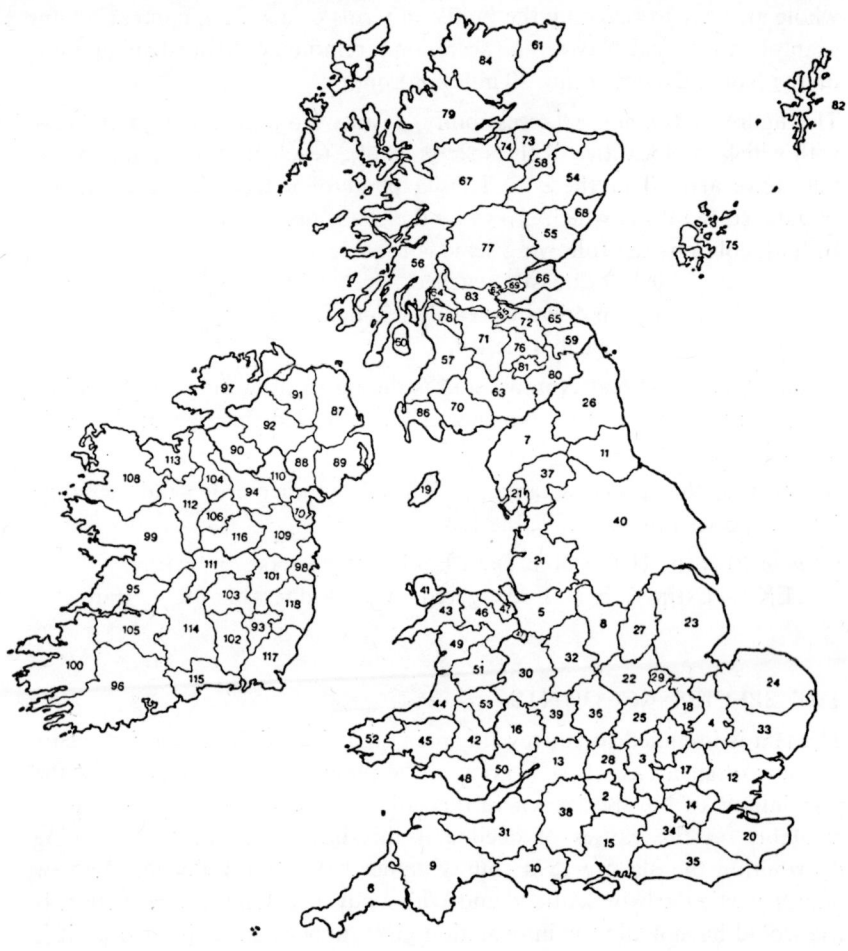

Figure 3
Pre-1974 county boundaries
Reproduced from Heath, J. (Ed.) 1976, *The Moths and Butterflies of Great Britain and Ireland*, London and Oxford: Curwen Press and Blackwell Scientific Publications with permission.

give information which in itself might result in a threat to the species. All the material collected by the Biological Records Centre during the survey is filed and kept securely at Monks Wood Experimental Station, where it is being continuously updated. Detailed information for any species or its sites is available to any bona fide member of a conservation organisation involved in action to protect a species with the exception of certain confidential records deposited with BRC only on the understanding that it will not be passed on without permission of the depositor.

England
1 Bedfordshire
2 Berkshire
3 Buckinghamshire
4 Cambridgeshire
5 Cheshire
6 Cornwall and Scilly
            Isles
7 Cumberland
8 Derbyshire
9 Devon
10 Dorset
11 Durham
12 Essex
13 Gloucestershire
14 Greater London
15 Hampshire and Isle
            of Wight
16 Herefordshire
17 Hertfordshire
18 Huntingdonshire
19 Isle of Man
20 Kent
21 Lancashire
22 Leicestershire
23 Lincolnshire
24 Norfolk
25 Northamptonshire
26 Northumberland
27 Nottinghamshire
28 Oxfordshire
29 Rutland
30 Shropshire
31 Somerset
32 Staffordshire
33 Suffolk
34 Surrey
35 Sussex
36 Warwickshire
37 Westmorland
38 Wiltshire
39 Worcestershire
40 Yorkshire

Wales
41 Anglesey
42 Breconshire
43 Caernarvonshire
44 Cardiganshire
45 Carmarthenshire
46 Denbighshire
47 Flintshire
48 Glamorgan
49 Merionethshire
50 Monmouthshire
51 Montgomeryshire
52 Pembrokeshire
53 Radnorshire

Scotland
54 Aberdeenshire
55 Angus
56 Argyll
57 Ayrshire
58 Banffshire
59 Berwickshire
60 Bute
61 Caithness
62 Clackmannanshire
63 Dumfriesshire
64 Dunbartonshire
65 East Lothian
66 Fifeshire
67 Inverness-shire
68 Kincardineshire
69 Kinross
70 Kirkcudbrightshire
71 Lanarkshire
72 Midlothian
73 Moray
74 Nairn
75 Orkney
76 Peebleshire
77 Perthshire
78 Renfrewshire
79 Ross and Cromarty
80 Roxburghshire

81 Selkirkshire
82 Shetland
83 Stirlingshire
84 Sutherland
85 West Lothian
86 Wigtownshire

Northern Ireland
87 Antrim
88 Armagh
89 Down
90 Fermanagh
91 Londonderry
92 Tyrone

Republic of Ireland
93 Carlow
94 Cavan
95 Clare
96 Cork
97 Donegal
98 Dublin
99 Galway
100 Kerry
101 Kildare
102 Kilkenny
103 Laios
104 Leitrim
105 Limerick
106 Longford
107 Louth
108 Mayo
109 Meath
110 Monaghan
111 Offaly
112 Roscommon
113 Sligo
114 Tipperary
115 Waterford
116 Westmeath
117 Wexford
118 Wicklow

THE BRITISH LIST IN A EUROPEAN CONTEXT

In 1974 the Council of Europe set up an ad hoc Group on Threatened or Endangered Plants in Europe, of which F. H. Perring was a member, to consider species rare or threatened on a continental scale. After an initial meeting to discuss the problem and devise a methodology the work of preparing a list was entrusted to the newly formed Threatened Plants Committee Secretariat of the International Union for Conservation of Nature and Natural Resources (IUCN). The Secretariat is based at the Royal Botanic Gardens, Kew.

The Secretariat proceeded with commendable speed, under the guidance of its Secretary, Mr. G. Ll. Lucas and the chairman of its European Sub-Committee, Dr. S. M. Walters. Through the networks set up to prepare *Flora Europaea* and *Atlas Florae Europaeae* they obtained comprehensive national lists from 15 countries as well as less precise information from elsewhere. As a result the Secretariat presented a Report to the Council of Europe in September 1976 — *List of rare, threatened and endemic plants for the countries of Europe.*

The list is in three sections. The first gives the full list of rare and threatened plants in Europe with their continental and world IUCN Red Data Categories. The second section is arranged by countries, and for each the list is in two parts: a) all species endemic to that country whether or not rare or threatened, and b) non-endemic rare and threatened species. For both a) and b) national and continental Red Data Categories are given. The third section is of non-endemic species still being considered as candidates for the list.

Whilst the European list was being prepared it became apparent that it would be to the advantage of that work and our own *Red Data Book* if there were the greatest possible co-ordination between the two publications. However the IUCN list differed from ours in three respects:

1. It included endemics whether threatened or not.
2. It included some non-endemic species on the list for the United Kingdom which are threatened in Europe as a whole but not in Britain.
3. It gave IUCN Red Data Categories.

These have been accommodated in this work as follows:

Endemics: Britain is not rich in endemic species of vascular plants. The IUCN list contains 15 names, not including *Alchemilla* (1) and *Sorbus* (13)*. Most of these 15 were included in our *Red Data Book* because of their rarity, however four species not in our list have been added:

> *Euphrasia marshallii*   *Primula scotica*
> *Gentianella anglica*   *Rhynchosinapis monensis*

Non-endemics: There are three species which are apparently declining so rapidly in Europe as a whole that they qualify for inclusion in the European list yet which are still frequent enough in

---

*France has over 70, Spain over 480 and Greece (including Crete and Aegean Islands) over 670.

Britain not to qualify for our list. They are:

Hammarbya paludosa   *Pilularia globulifera*
Petroselinum segetum

These four endemic and three non-endemic species are included in the species accounts but appear in square brackets, and they are not included in Table 2. Though they are not rare or threatened in Britain we clearly hold the total populations or a very large proportion of them for all seven species and our conservation policies should pay particular attention to their survival in this country.

IUCN Red Data Book Categories: In order to provide the Threatened Plants Committee with the information they required all the plants in the British list had to be given an IUCN Category. For completeness and to further integrate the two works this Category for Britain has been included in the species accounts, where it appears to the right of the Threat Number (TN), and in Appendix I and the Index. For British plants the relevant Red Data Book Categories are as follows:

Ex *Extinct.*

E *Endangered.*
Taxa in danger of extinction and whose survival is unlikely if the causal factors continue operating.
Included are taxa whose numbers have been reduced to a critical level or whose habitats have been so drastically reduced that they are deemed to be in immediate danger of extinction.

V *Vulnerable.*
Taxa believed likely to move into the endangered category in the near future if the causal factors continue operating.
Included are taxa of which most or all the populations are *decreasing* because of over-exploitation, extensive destruction of habitat or other environmental disturbance; taxa with populations that have been seriously *depleted* and whose ultimate security is not yet assured; and taxa with populations that are still abundant but are *under threat* from serious adverse factors throughout their range.

R *Rare.*
Taxa with small populations that are not at present endangered or vulnerable, but are at risk.
These taxa are usually localized within restricted geographical areas or habitats or are thinly scattered over a more extensive range.

N.B. In practice, Endangered and Vulnerable categories may include, temporarily, taxa whose populations are beginning to recover as a result of remedial action, but whose recovery is insufficient to justify their transfer to another category.

Though the IUCN Categories are determined in a different manner from the Threat Number they are based on the same data and there is a strong correlation between the two. However it is not absolute. In general species with TN = 7 or less will only be IUCN Category R whilst species with TN = 10 or more will be in Category V or E: those with TN = 8 or 9 may be in any category.

Apart from the endemics mentioned above only 14 species which are in the British list appear in the IUCN European List. In the text their European Red Data Category is given in addition to the Category for Britain. These species are:

Apium repens   *Limonium transwallianum*
Carex recta   *Liparis loeselii*
Cypripedium calceolus   *Potamogeton epihydrus*
Dianthus gratianopolitanus *Rumex rupestris*
Galium fleurotii   *Senecio congestus*
Gentianella uliginosa   *Spiranthes romanzoffiana*
Limonium paradoxum   *Trichomanes speciosum*

The inclusion of these species in the European list adds to the importance of ensuring that they are adequately protected in this country, though we are too late to save *Senecio congestus* which became extinct about 1900.

CAUSES OF DECLINE

It is not possible at this time to make a complete analysis of the many factors which threaten the flora of Great Britain. However some indication of the more important factors may be derived from Table 2. To compile it the 321 rare species included in this book have been allocated to one of nine broad habitat classifications. Within each habitat class they have been placed in the IUCN Red Data Category: Extinct, Endangered, Vulnerable and Rare. Their distribution in these four categories is a measure of the extent to which the rare species of that habitat are threatened.

The most severely threatened group of plants in the flora are the arable weeds: 96 per cent of the 23 species are threatened in some way and though only one species (*Bromus interruptus*) has so far become extinct 14 species (62 per cent) seem likely to become extinct in the near future. Species of other open man-made habitats — roadsides, quarries and so on, are also under pressure: 16 per cent of the 26 species in this category are already extinct, the highest percentage for any habitat whilst another 46 per cent

(12 species) of this group are threatened to some degree. Similar data from Belgium based on Delvosalle *et al.* (1969) and Van Rompaey and Delvosalle (1972) show that these two habitat groups combined, arable and open man-made habitats, are the most threatened in that country: 30 per cent (17 species) are already extinct whilst a further 66 per cent (38 species) are endangered or vulnerable.

The second group of plants most threatened in Great Britain are the wetland species: plants of bog, marsh and the margins of ponds and lakes.

Nearly one third (6) of the 19 species which have become extinct in this country are in this group and a further seven species are endangered. In contrast losses of submerged, aquatic species have so far been few: of the 12 species in this group only one (*Hydrilla verticillata*) has become extinct, none is endangered and only two are vulnerable.

Only one other group is less threatened than the aquatic species: that is the montane element. It is the second biggest group in our rare flora with 60 species (19 per cent) but the great majority of these (80 per cent) are rare species of Scottish mountains where, although threats from ski lifts and over-collecting operate locally most sites are adequately protected by their remoteness.

Thus it is clear that the main conservation problem of rare plants lies in the lowlands rather than in the Highlands of Scotland. 35 species of lowland grassland and other natural open habitats alone are at present threatened and the total number of lowland species in the Endangered or Vulnerable Categories is about 120.

Whilst National and other Nature Reserves have made a significant contribution to the conservation of rare plants the data so far collected shows that no more than 50 per cent of other sites have even the minimum protection of being declared Sites of Special Scientific Interest (SSSI): and 68 species marked 'No conservation' in the species accounts have no sites which are formally protected in any way, though 18 of these are arable weeds where such protection is not easily arranged. Nevertheless there is an urgent need to ensure that wherever possible at least the sites of all the Endangered and Vulnerable species are protected.

So far, in Great Britain, the number of extinctions in our flora since recording began in the seventeenth century has been small: only 19 species, which represents less than 1 per cent. However what is disturbing is the large number of Endangered species (46) which could soon become Extinct if we do not take adequate action. That action needs to be taken now. It is a formidable but not impossible challenge which it is hoped will be tackled jointly by all the wild life conservation organisations in this country.

LEGISLATION TO CONSERVE WILD PLANTS

When work began on this *Red Data Book* legislation for the protection of plants scarcely existed. There was a bye-law in most counties which made it an offence to dig up plants in places to which the public had access, but these had been forgotten and it is doubtful whether any prosecution had ever been brought. Other organisations like the National Trust and the Nature Conservancy had powers to apply bye-laws to their own properties, but, as indicated earlier, many rare and endangered species are not on reserves, or on National Trust property and, as the law stood, there was nothing to prevent the public from digging them up and destroying their sites for ever. However, after almost 12 years of sustained effort by the Society for the Promotion of Nature Conservation, the Council for Nature and the Botanical Society of the British Isles, who had formed between them the Wild Plants Protection Working Party (WPPWP), a Bill giving a measure of protection to wild plants received the Royal assent and became an Act of Parliament on 1 August, 1975. This Act entitled 'The Conservation of Wild Creatures and Wild Plants Act, 1975' combined two Bills, one on Wild Creatures and one on Wild Plants, presented to the House of Lords by the Earl of Cranbrook and Lord Beaumont of Whitley respectively. These two Bills were seen through the House of Commons in their combined form by Mr. Peter Hardy, M.P. who drew first place in the ballot for Private Members' bills in the 1974-75 session of Parliament.

The major clauses relevant to wild plants are as follows:

> If, save as may be permitted by or under this Act, any person other than an authorised person without reasonable excuse uproots any plant, he shall be guilty of an offence.

*　　*　　*

> If, save as may be permitted by or under this Act, any person without reasonable excuse picks, uproots of destroys any protected plant listed in Schedule 2, he shall be guilty of an offence unless the picking, uprooting or destruction occurs as an incidental result, which could not reasonably have been avoided, of any operation which was carried out in accordance with agricultural or forestry practice.

*　　*　　*

> A person shall not be guilty of an offence against this Act by reason only of the doing of anything in pursuance or furtherance of any obligation imposed, or in exercise of any powers conferred, by or under an Act of Parliament.

*　　*　　*

Where, on a representation made to him by the Nature Conservancy Council it appears to the Secretary of State to be necessary in the interest of the proper conservation of plants he may by order add any plant to, or remove any plant from, Schedule 2 to this Act.

An order made under this subsection may apply:

(a) to the whole or to particular provisions of this Act;
(b) generally or to a particular area;
(c) to plants in a particular category; or
(d) at all times or at particular times of the year;

and the order may make different provision for different circumstances.

*   *   *

A licence may be granted to any person by the appropriate authority stipulated in the next subsection authorising that person, notwithstanding anything in this Act or in any order made under this Act, but subject to compliance with any specified conditions for scientific or educational purposes or for the conservation of plants to pick or uproot within a specified area by any specified means any plant of a specified species.

The appropriate authority for the grant of a licence shall be the Nature Conservancy Council.

*   *   *

Any person guilty of an offence under this Act shall be liable on summary conviction to a fine not exceeding £100. Provided that, where the offence was committed in respect of more than one species of plant, the maximum fine which may be imposed under this subsection shall be determined as if the person convicted had been convicted of a separate offence in respect of each protected species of plant.

*   *   *

The Nature Conservancy Council at any time may, and five years after the passing of this Act and every five years thereafter shall, review the Schedules to this Act and shall advise the Secretary of State if any plant has become so rare that its status as a British plant is being endangered by any action designated as an offence under this Act and it should be included in Schedule 2 either generally or with respect to a particular area or in relation to a particular category and either at all times or at particular times of the year, or has become so common that its status is no longer endangered and it should be removed therefrom.

*   *   *

A local authority shall take such steps as they consider expedient for bringing the effect of this Act to the attention of the public and in particular schoolchildren.

*   *   *

A local authority in England or Wales shall have the power to institute proceedings for any offence under this Act committed within their area.

Species of Protected Plants

Common name	Scientific name
Alpine Gentian	*Gentiana nivalis*
Alpine Sow-thistle	*Cicerbita alpina*
Alpine Woodsia	*Woodsia alpina*
Blue Heath	*Phyllodoce caerulea*
Cheddar Pink	*Dianthus gratianopolitanus*
Diapensia	*Diapensia lapponica*
Drooping Saxifrage	*Saxifraga cernua*
Ghost Orchid	*Epipogium aphyllum*
Killarney Fern	*Trichomanes speciosum*
Lady's-slipper	*Cypripedium calceolus*
Mezereon	*Daphne mezereum*
Military Orchid	*Orchis militaris*
Monkey Orchid	*Orchis simia*
Oblong Woodsia	*Woodsia ilvensis*
Red Helleborine	*Cephalanthera rubra*
Snowdon Lily	*Lloydia serotina*
Spiked Speedwell	*Veronica spicata*
Spring Gentian	*Gentiana verna*
Teesdale Sandwort	*Minuartia stricta*
Tufted Saxifrage	*Saxifraga cespitosa*
Wild Gladiolus	*Gladiolus illyricus*

This Act, which covers the whole of Great Britain means that it is now an offence, with a fine of up to £100 upon conviction, for anyone without permission of the owner, or occupier of the land or their agent, to dig up plants in the countryside. A further offence is committed if any of 21 species on Schedule 2 is picked, uprooted or destroyed.

However, the Nature Conservancy Council can issue licences for scientific, educational or conservation purposes which would make digging up of particular species or collecting of scheduled plants lawful.

Comparison of the list of protected plants with the Threat Numbers in this book will show that the list does not contain all the most highly threatened species, whilst others which are included have quite low Threat Numbers.

This is because historical and political considerations had to be balanced with, and at times outweighed, strictly scientific ones during the passage of the Bill. However the Act contains provisions for the alteration of the Schedule by the Secretary of State on advice from the Nature Conservancy Council, who must review the Schedule at least once every five years. In future purely scientific consideration will doubtless be given greater priority.

THE FUTURE

No work of this kind can ever be definitive. Change is continuous and whilst this book is being printed some sites may be lost and others refound. We hope that for some species reported as extinct extant populations will be found: certainly experience suggests the surest way to 'revive' a species is to publish a statement of its extinction. Other species, notably arable weeds and wetland species, are declining rapidly and some could already be approaching the criteria for inclusion.

For this reason we believe that this should be only the first of a series of *Red Data Books* on the British Vascular Flora. Ideally a revised version should appear every five years in relation to the quinquennial review of the Schedule by the NCC required by the Act.

But change is not the sole reason for revision. In many senses this work is still incomplete:

1. Many of the species in this book, particularly in the remote areas, were not visited during the survey. We need to know whether they still exist and to have population forms for those which are refound.

2. Even for sites for which recent records have been made there are still over 1,000 lacking population forms — these sites need to be revisited and the essential work completed.

3. The 321 species herein listed do not include all those in the British vascular flora which may be endangered. As explained on p. vi two major critical genera, *Hieracium* and *Rubus* have been omitted. Taxonomic treatment of the latter genus is still unsatisfactory but recent accounts of *Hieracium* by Sell & West (1968, 1976) and the knowledge of the distribution of the 250 species which occur in the British Isles which they have assembled, which was published in Perring (1968) and the original data which are stored at BRC, is the basis for beginning a survey of the rare species. This genus is particularly important because one half of the taxa are endemic.

4. No attempt has yet been made, with certain known exceptions, to include inter- and infra-specific taxa. The recent work by Stace (1975) describing nearly 1,000 hybrids in our flora is an indication of the size of the inter-specific problem. The number of infra-specific taxa, many undoubtedly endemic, remains unknown but will emerge during the next decade if, as is hoped, work on a new critical Flora of the British Isles progresses. The sites of many of these taxa will undoubtedly need investigation.

As new information in all the above categories becomes available it will have to be assembled and assessed and incorporated in future editions of this work.

But survey alone is not enough. We need to know much more about the biology of each species. The population form is a good beginning and at BRC we are building up a collection of published and unpublished papers and notes, and a collection of habitat photographs, but if the action necessary for the conservation of our rare species is to be fully effective it needs to be based on an autecological study of the species in this country, and, possibly, elsewhere in Europe. The staff in BRC will never be able to investigate more than a very few of these though one of us (L.F.) has already begun work on *Orchis militaris*. Until now only 16 species in this *Red Data Book* have been the subject of accounts in the *Biological Flora of the British Isles* in the *Journal of Ecology*. Here is still a vast and vitally important area of research to be undertaken.

Because of the long tradition of interest in our flora which still continues amongst the 2,000 + members of the BSBI and because of the well developed network of conservation organisations both official and voluntary we have an opportunity to ensure that the flora of our islands is better protected than that in other parts of the world. Perhaps largely through ignorance too many species have reached the point where they are endangered. If this book serves to break down that ignorance, create an awareness of the size of the problem which exists and stimulate the wide scale action that is necessary then the staff of the BRC and all those who have been involved in its preparation will be handsomely rewarded.

ACKNOWLEDGMENTS

This volume could not have been prepared without the selfless assistance of British field botanists too numerous to mention, and in particular the County Recorders of the Botanical Society of the British Isles. Credit for beginning to assemble the data is largely due to Miss M. N. Hamilton our colleague in the institute now at our Merlewood Research Station. We are grateful too to Mrs G. Crompton, R. S. R. Fitter, Dr. S. W. Greene, the late J. E. Lousley, D. McClintock, R. Mackechnie, E. C. Wallace and Dr. S. M. Walters, who read much of the text in draft and made many useful comments, but they must in no way be held responsible for any errors which have been included — any criticism properly falls on us. Special mention must be made of Mrs. C. Shackcloth ('Connie') who prepared the typescript from the almost impossible handwriting of one of us (F.H.P.) and yet remained cheerful to the end.

Finally special thanks are due to fellow members of the Conservation Liaison Committee of the SPNC for their invaluable advice and their incredible tolerance in waiting eight years for the completion of this work with their customary patience and good humour, especially to Tim Sands who has helped in so many ways in ensuring the smooth production of the book.

REFERENCES

Dandy, J. E. 1958. *List of British Vascular Plants*. London: British Museum.
Dandy, J. E. 1969. Nomenclatural changes in the 'List of British Vascular Plants'. *Watsonia*, 7, 159-78.
Delvosalle, L., Demaret, F., Lambinon, J. and Lawalrée, A. 1969. *Plantes rares, disparues ou menacées de disparition en Belgique: L'appauvrissement de la flora indegène*. Service dés Reserves Naturelles domaniales et de la Conservation de la Nature, Travaux No. 4.
Dony, J. G., Perring, F. H. and Rob, C. M. 1974. *English Names of Wild Flowers*. London: Butterworths.
Lucas, G. Ll. and Walters, S. M. 1976. *List of Rare, Threatened and Endemic Plants for the Countries of Europe*. Kew: IUCN Threatened Plants Committee.
Perring, F. H. (Ed.) 1968. *Critical Supplement to the Atlas of the British Flora*. London: B.S.B.I./Nelson.
Perring, F. H. (Ed.) 1970. *The Flora of a Changing Britain*. B.S.B.I. Conference Report No. 11.
Perring, F. H. and Walters, S. M. (Eds.) 1962. *Atlas of the British Flora;* London: B.S.B.I./Nelson. 2nd Edn. 1976. Wakefield: B.S.B.I./EP Publishing.
Sell, P. D. and West, C. 1968. *Hieracium*. In Perring, F. H. 1968, 75-134.
Sell, P. D. and West, C. 1976. *Hieracium*. In Tutin, T. G. *et alia* 1976. *Flora Europaea*, 4, 358-410.
Stace, C. A. (Ed.) 1975. *Hybridization and the Flora of the British Isles*. London: Academic Press.
Tutin, T. G. *et alia* 1964-76. *Flora Europaea* Vols. 1-4. London: Cambridge University Press.
van Rompaey, E. and Delvosalle, L. 1972. *Atlas de la Flore Belge et Luxembourgeoise*. Bruxelles: Jardin Botanique National.

EXPLANATION OF THE TEXT

Nomenclature

The scientific name is based on two works: *List of British Vascular Plants* by J. E. Dandy (1958) and *Nomenclatural changes in the List of British Vascular Plants* by J. E. Dandy (1969). In addition name changes published in *Flora Europaea* (T. G. Tutin *et al.* 1964 *et seq.*) based on nomenclatural research have been included.

The English name, where given, is taken from *English Names of Wild Flowers* by J. G. Dony, C. M. Rob and F. H. Perring (1974).

Present and past distribution

GB = Great Britain (including off-shore islands)
S = Channel Isles (Sarnia)
H = Ireland (Hibernia) including the Republic and N. Ireland.

Conservation status

NNR = National Nature Reserve
NCTR = Nature Conservation Trust Reserve
FNR = Forest Nature Reserve
LNR = Local Nature Reserve
FSCR = Field Studies Council Reserve
RSPBR = Royal Society for the Protection of Birds Reserve
NT = National Trust property
SSSI = Site of Special Scientific Interest

The figure refers not to the number of reserves etc. on which the species occurs but to the number of localities (see p. xii).

Threat

TN = Threat number (see p. x)
E = Endangered, V = Vulnerable, R = Rare (see p. xvii)
IUCN List = Species in the *List of Rare, Threatened and Endemic Plants for the countries of Europe* (see p. xvii), but not rare in GB.

Scheduled Species

All species thus marked are species totally protected by the Conservation of Wild Creatures and Wild Plants Act, 1975. (see p. xxii).

ACCOUNTS OF INDIVIDUAL SPECIES

Isoetes histrix Bory   Land Quillwort

At least seven colonies of this terrestrial species, one of which is much smaller than the others, are known in Cornwall where it grows in moist maritime turf; it also occurs, presumably in the same habitat, on two of the Channel Isles. There appears to be no immediate threat to the plant; it is not at risk from collectors as it is invisible above ground from April to October, and no special conservation measures are required.

2/2 GB post 1960   2 SSSI   TN = 6 R
2/2   S post 1970

Equisetum ramosissimum Desf.

The species, which is doubtfully native, occurs in only one riverside locality in Lincolnshire, on a clay bank formed when the river was straightened in 1880-87. The colony extends over 200 metres but has recently been threatened by mowing. It could also be endangered by further straightening of the river, or by expansion of a large rubbish tip just behind the bank.

1/1 GB post 1970   No conservation   TN = 9 E

Trichomanes speciosum Willd.   Killarney Fern

Although now recorded from only a few localities in S.W. and N.W. England and Wales this fern is still most plentiful in S.W. Ireland where, early in the nineteenth century, it occured in profusion, subsequently being reduced considerably by the depredations of botanists, gardeners, dealers and tourists so that it continued to exist in only a few inaccessible places amongst rocks, in deep shade and high humidity. The exact number of extant stations in Britain is not known, but most populations are very small, at least one being near extinction, primarily because of collecting. However, at one British station 200-300 fronds were counted recently, and in another about 1000.

 8/15 GB post 1950   2 NNR : 1 FNR : 1 LNR   TN = 9 V
22/47 H post 1930   Scheduled species   V in Europe

Cystopteris dickieana Sim

Though now lost from its two Perthshire stations, this fern is still present in a sea-cave in Kincardineshire where, over the past 15 years, numerous plants have been recorded, though the strength of the population appears to fluctuate with natural changes in the cave. Being a taxonomically difficult species, it may subsequently be found to be more widely distributed; recently, a phenotypically identical specimen was collected in Cornwall.

1/3 GB post 1970   1 SSSI   TN = 10 E

1

Woodsia ilvensis

Woodsia ilvensis (L.) R. Br.   Oblong Woodsia

This montane fern of rock-crevices is now known from about four localities in Caernarvonshire, Cumberland and Dumfriesshire; formerly it was also recorded in Merionethshire, Durham, Westmorland, Angus and, possibly in error, from Perthshire and the Inner Hebrides. Whilst most colonies consist of 3-20 plants, a fine one is known in Lakeland comprising, in 1965, over 100 plants, some being luxuriant. A decline has occurred because of over-collecting in the past; this threat continues.

4/12 GB post 1970   1 NNR : 3 SSSI   TN = 9 V
                           Scheduled species

Woodsia alpina (Bolton) S.F. Gray   Alpine Woodsia

This fern is found at altitudes of up to 3000 ft. (920 m.), growing in rock crevices; at present at least 15 localities are known in Caernarvonshire, Perthshire, Angus, Inverness-shire and Argyllshire, in some of which there are good colonies — 24 tufts at one station in Perthshire and over 100 in another, whilst the situation in Wales is reasonably promising. Much collecting in the past has caused reduction and this has also happened recently in at least two stations; the species is now confined to the more inaccessible parts of its range.

11/17 GB post 1950   5 NNR   TN = 7 R
                      Scheduled species

Dryopteris cristata (L.) A. Gray   Crested Buckler-fern

A fern of wet heaths, fens, marshes and dunes, this is now known from about 21 stations in Surrey, Suffolk, Norfolk and Renfrewshire. In Surrey it is threatened by the recutting of a drain and the planting of conifers; in Renfrewshire the loch-margin community of which it is part is vulnerable to wave action and to trampling. In Norfolk it is not infrequent but is decreasing locally. It is thought to be extinct in old localities in Kent, Huntingdonshire, Staffordshire, Nottinghamshire, Cheshire and Yorkshire.

9/26 GB post 1960   1 NNR : 1 NT : 3 SSSI   TN = 8 V

Pilularia globulifera L.   Pillwort

This creeping rhizomatous fern occurs in shallow water at the margin of acid ponds and lakes. It is widespread in Great Britain where it has been recorded from 69 vice-counties, but is much more restricted in Ireland where it has been found in only ten, mainly along the west coast. It is a species which has declined considerably with the filling of ponds and drainage and now appears to be extinct in over 40 of these vice-counties, as well as in its single locality in Jersey in the Channel Isles. Whilst it is still known from about 70 localities in Great Britain it is not yet threatened. However it is threatened in Europe as a whole and our populations are therefore of some significance.   IUCN List, V in Europe

Ophioglossum lusitanicum L.

This Mediterranean fern occurs only in the Isles of Scilly and on Guernsey. The main colony in Scilly consists of about 100 plants over an area of one square metre. Because the fronds are active in the winter and disintegrate by early April, the species is often overlooked.

1/1 GB post 1960   1 SSSI   TN = 7 R
1/1   S post 1970

Ranunculus ophioglossifolius Vill.   Adder's-tongue Spearwort

Until recently this annual was thought to occur in only one station in Gloucestershire where active management of the habitat is carried out by the Nature Conservation Trust; the population, which varies in strength from year to year, is now almost certainly safe, though *Glyceria* is a rampant competitor. Now, however, a second colony, previously believed to be extinct, is thought to exist, also in Gloucestershire. Formerly the species was recorded from Dorset, Hampshire and Jersey but is now extinct in all of these.

2/4 GB post 1970   1 NCTR   TN = 9 E
0/1   S post 1928

Paeonia mascula (L.) Mill.   Peony

The only station is on the cliffs of the island of Steep Holme where this perennial is known to have been established since at least 1803; then abundant, it had decreased considerably by the end of the nineteenth century because of uprooting by tourists so that, on two separate occasions, only two plants were seen in inaccessible positions. Recently fewer than twenty plants were recorded, then all in fruit. The island is now administered by a Trust, and more adequate protection is given. Formerly a plant thought to be this species was also recorded from Somerset where, however, it may have been a garden outcast.

1/2 GB post 1960   Private Trust   TN = 9 V

Fumaria occidentalis

Fumaria occidentalis Pugsl.

An endemic annual of arable fields and waste places, abundant in at least 20 stations in Cornwall and the Isles of Scilly. It is plentiful in bulb-fields in Scilly where the known localities are scattered over an area of approximately one square kilometre and the species occurs in about one tenth of this. It appears to have increased slowly during the last 40 years. On the mainland it grows by a railway, in lanesides and on waste ground where it is in need of adequate protection. No threat is known and the species appears to be immune to the effects of chemical sprays, at least in Scilly.

11/12 GB post 1950   1 NT : 1 SSSI   TN = 6 R

Fumaria martinii Clavaud

This very rare annual of arable land has been recorded in about ten stations in Cornwall, Devon, Somerset, Sussex, Surrey and Guernsey in the Channel Isles in the period 1928-1962, but it is doubtful whether it exists in any of these now. Visitors to several of these former localities since 1970 have been unable to refind this plant. It is known that in Cornwall at least one population has disappeared since 1955 because the habitat has been replaced by grassland, and the same fate befell the single Surrey locality.

0/10 GB post 1962   No conservation   **? EXTINCT**
0/1   S post 1928

Rhynchosinapis monensis (L.) Dandy   Isle of Man Cabbage

This endemic biennial still occurs in at least 22 localities in sandy areas close to the sea in Glamorgan, Carmarthenshire, Lancashire, Cumberland, Isle of Man, Kirkcudbrightshire, Ayrshire and on the Isle of Arran. Perhaps less frequent than formerly around the mouth of the River Clyde.

                                              IUCN List

Rhynchosinapis wrightii (O.E. Schulz) Dandy   Lundy Cabbage

This endemic is a short-lived perennial known only from granite and slate cliffs and slopes on the east side of Lundy Island where its relative inaccessibility provides some protection. However threats exist from grazing by goats, sheep and deer and from growing tourist pressure in the part of the island where it is most abundant.

1/1 GB post 1970   Private Trust   TN = 5 R

Isatis tinctoria L.   Woad

Introduced for dyeing more than a thousand years ago, this biennial or perennial herb has only persisted in a naturalised state, in one locality on a riverside cliff in Gloucestershire where it has been known since 1818 and in Surrey, where it has been known since before 1800. In the latter site it is now protected by the owners of the modern houses built below the chalk cliffs on which it grows. Elsewhere, as a casual or relic of cultivation, it was recorded before 1930, in 29 vice-counties mainly in England but also in Scotland and Ireland. Although the Gloucestershire site is protected the plant is endangered by natural erosion accentuated by public pressure.

2/2 GB post 1960   1 NCTR   TN = 9 V

Thlaspi perfoliatum L.   Perfoliate Penny-cress

This overwintering annual is confined to bare ground, spoil heaps, walls and similar habitats, on limestone and is native only in Wiltshire, Oxford-shire and Gloucestershire. Altogether, it is now known in 11 stations at one of which, in Oxfordshire however, it is threatened by clearing operations, and in another by the dumping of rubbish, though it is plentiful in a third station on a railway bank. In Gloucestershire it appears to be spreading for new stations have been recorded recently. It has also spread to, and become established in, several other places in southern England.

7/14 GB post 1950   1 NCTR   TN = 7 R

Cochlearia micacea E.S. Marshall

This biennial to perennial herb is confined to a few high mountains in Perthshire, Angus and Ross. It is probably endemic to Scotland. It has only recently been re-assessed as a distinct taxon and little is known of the present size of the populations.

3/3 GB post 1970   2 NNR   TN = 6 R

Alyssum alyssoides (L.) L.   Small Alison

This rare annual, an alien of grassy fields and arable land, has been recorded from many locations throughout Great Britain, most of the records having been made before 1930, and only nine having been con-firmed since 1950. In most of these localities it was merely a casual, but it persists from year to year in the Breckland of Suffolk and Sandlings of Norfolk. It has also been recorded in Ireland but has never become established there.

2/6 GB post 1950   No conservation   TN = 13 E

Draba aizoides L. Yellow Whitlowgrass

This tufted perennial occurs only in Glamorgan where it is locally abundant on the turf, or in crevices of coastal limestone cliffs, over a distance of about ten miles (16 km.), and on old ruins and walls in the neighbourhood; populations range in size from isolated individuals to a few hundred plants. When first recorded, the plant was generally abundant, since when the distribution does not seem to have changed, except by its disappearance from some of the more accessible sites through depredation by gardeners and other collectors; it is naturally protected to some extent by its early flowering period and mainly inaccessible habitat, though sheep graze the young inflorescences in early spring. Sheep and man may cause much damage by trampling outside the National Nature Reserves by which it is partially protected.

3/3 GB post 1960   2 NNR : 1 SSSI   TN = 5 R

Arabis alpina L. Alpine Rock-cress

Only two small colonies of this perennial mat-forming herb are known at about 2700 ft. (820 m.) on wet rock ledges in the mountains of Skye; rumours of other stations have not been confirmed. The sites are on private land and are far enough off the main climbing routes to be reasonably safe.

1/1 GB post 1960   1 SSSI   TN = 7 R

Arabis scabra All. Bristol Rock-cress
A. stricta Huds.

This perennial is confined to the Carboniferous limestone rocks of the Avon Gorge, both in Somerset, where only a small number were observed in 1971, and in Gloucestershire, where thousands of plants were recorded. Formerly, it was found at many spots within this limited area, on loose rubble and in turf, as well as on the live rock but, because the locality is freely accessible to the public and the plant is somewhat conspicuous, it has declined and will continue to do so as pressure increases. An old record from Radnorshire remains unconfirmed and it is extinct at a station in Somerset where it was introduced and naturalised.

1/1 GB post 1960   1 NNR   TN = 9 V

Rorippa austriaca (Crantz) Bess. Austrian Yellow-cress

An alien occasionally established in ditches and on river and railway banks.

Now known only from Surrey, Middlesex, Berkshire, Oxfordshire, Gloucestershire, Monmouthshire, Glamorgan, Leicestershire, Nottinghamshire and Lancashire. Previously recorded from Kent, the Huntingdon/Northampton border area and from Limerick in Ireland.

13/15 GB post 1950   No conservation   TN = 7 R
 0/1   H post 1930

Matthiola incana (L.) R.Br.   Hoary Stock

This annual to perennial herb of sea cliffs is doubtfully native, but apparently abundant, in inaccessible situations on the Isle of Wight and in Sussex, and is known as a certain introduction in about 15 other vice-counties including the Channel Isles where it was first recorded in 1578.

3/4 GB post 1950 (as possible native)   3 SSSI   TN = 6 R

Matthiola sinuata (L.) R.Br.   Sea Stock

Now only in Devon, Glamorgan and the Channel Isles, this biennial of sea cliffs and dunes was previously also recorded from Pembrokeshire, Merionethshire, Caernarvonshire, Anglesey and Ireland, in most of which it was extinct by the beginning of the nineteenth century. In one station in Devon it is thriving and increasing and, in another, occasional on stabilised dunes; in its sole station in Glamorgan it is abundant. The reason for its disappearance is unknown but it was possibly over-collected, as in Jersey now, or else the habitat may have been altered or destroyed. The record for the Isles of Scilly is now considered to have been an error.

2/15 GB post 1970   1 NNR : 2 SSSI   TN = 11 V
2/2   S post 1970
0/8   H post 1930

Sisymbrium irio L.   London Rocket

This overwintering annual which became abundant on the ruins of London after the great fire of 1666 subsequently became established on roadsides, on walls and in waste places in towns and cities scattered throughout Great Britain and in Dublin in Ireland. Once recorded from 21 vice-counties since 1960 it has only been reported from Devon, Kent, Middlesex, Suffolk, Lincolnshire, Lancashire and Selkirk, and it only persists in three or four sites in London.

14/48 GB post 1960   No conservation   TN = 10 V
 2/5   H post 1930

7

Viola rupestris

Viola rupestris Schmidt   Teesdale Violet

This perennial is known from 11 localities on open, mossy, sheep-grazed turf or bare ground on limestone in Yorkshire, Durham and Westmorland. One of the populations extends over several hundred metres but one of the others covers only 10 square metres. Though it appears to be adequately protected now, threats exist from collectors, because of its rarity, the proximity of an easy access road in one area and from the planting of conifers.

6/6 GB post 1960   1 NNR : 2 SSSI   TN = 6 R

Viola persicifolia Schreb.   Fen Violet
V. stagnina Kit.

In England this is a fen species but, in Ireland it is found in damp grassy hollows on limestone. Following the destruction of habitats, through drainage, in Suffolk, Norfolk, Cambridgeshire, Nottinghamshire and Yorkshire, it has disappeared from about 18 known stations in these counties, but it still occurs in one locality in Huntingdonshire and another in Yorkshire. It has apparently been lost through hybridisation in Oxfordshire. In the best-known site, in Huntingdonshire, there were thousands of plants in 1975, mainly on newly disturbed peat, but also on mown rides and it is dependent on such disturbance for survival; in 1972 only about 50 plants were observed. It has not declined so rapidly in Ireland, though few details are known of the current situation there.

3/17 GB post 1960   1 NNR : 1 SSSI   TN = 10 E
7/11 H post 1930

Viola kitaibeliana Schult.   Dwarf Pansy

This erect annual of coastal sand dunes, dry grassy places and cliffs, occurs only in the Isles of Scilly where it grows on five islands in small quantity and in the Channel Isles, where it is more plentiful. It appears to be frequent to abundant, and will be endangered only if digging for sand ceases and rabbits become extinct!

1/1 GB post 1970   1 SSSI   TN = 8 R
3/3   S post 1960

Polygala amara L.   Dwarf Milkwort

This creeping perennial is known in at least 13 stations in limestone grassland and damp mountain pastures in Yorkshire and Durham. It was formerly somewhat more abundant, but the reason for its decline is not known.

6/6 GB post 1960   2 NNR : 1 NCTR : 2 SSSI   TN = 6 R

Polygala austriaca Crantz

Confined to Kent, where it occurs in 13 stations in open grassy spots on chalk downs; at least one population is now extinct through ploughing and overgrowth of rough herbage; another is small because of heavy grazing; the others are large to very large. Formerly also in Surrey, but now extinct there.

7/9 GB post 1960   2 NNR : 1 NCTR : 5 SSSI   TN = 6 R

Hypericum linarifolium Vahl   Flax-leaved St. John's-wort

A perennial of dry rocky slopes on acid soils, this is now certainly only in Devon, where at least five localities are known, in two of which it is reasonably abundant, and in Caernarvonshire, where over 100 plants occur. In the Channel Isles the species is locally plentiful on dry, heathy hillsides over granite in Jersey; it has been recorded occasionally on Alderney, but only a few plants are known from Guernsey. Formerly it was also recorded from Cornwall, Radnorshire, Merionethshire and Anglesey. The reason for its decline is uncertain though, in Cornwall, this may have been due to deliberate collecting. It was recorded once as a casual in Shropshire.

4/10 GB post 1950   No conservation   TN = 10 V
3/4   S post 1960

Tuberaria guttata (L.) Fourr.   Spotted Rock-rose
    subsp. **breweri** (Planch.) E.F. Warb.

This endemic subspecies is an annual which occurs on exposed rocky moorland near the sea; now at five stations in Anglesey, two of which are known to support large and healthy populations, and one in Caernarvonshire. It still occurs in several localities in the west of Ireland.

4/5 GB post 1970   1 SSSI   TN = 10 V
4/7 H post 1950

Helianthemum apenninum (L.) Mill.   White Rock-rose

This perennial occurs on limestone rocks in Devon and Somerset where it is abundant in at least five of its eight known stations. No threat appears to exist as most stations are protected to some degree.

4/4 GB post 1970   1 LNR : 3 NT   TN = 8 R

Helianthemum canum

Helianthemum canum (L.) Baumg.   Hoary Rock-rose
   subsp. **levigatum** M.C.F. Proctor

The only locality for this endemic perennial is in Yorkshire where the smallish population is restricted to a few outcrops of metamorphic sugar-limestone near the bleak and exposed fell summit.

1/1 GB post 1970   1 NNR   TN = 6 R

Elatine hydropiper L.   Eight-stamened Waterwort

This annual of ponds, small lakes and canals is now known from Worcestershire where, until recently, it had been believed extinct, Anglesey, Renfrewshire, Stirlingshire, Dunbartonshire and Perthshire. Formerly it was also in Sussex, where it was last seen in 1960, Surrey, Berkshire and Staffordshire: extinction of at least some colonies has been due to an alteration in the water level which was also the cause of its temporary disappearance (1962-1968) from Worcestershire. The first records for Scotland were made in 1968, where two populations of considerable size are now known. In Ireland, since 1930, it has been recorded from Armagh, Down and Antrim, most recently from L. Neagh in 1971.

 6/15 GB post 1960   4 NNR : 1 RSPBR : 1 SSSI   TN = 5 R
11/14 H post 1930

Silene otites (L.) Wibel   Spanish Catchfly

This perennial is confined to the Breckland heaths of Suffolk, Norfolk and Cambridgeshire, being widespread over this area; occasionally recorded as a casual elsewhere. In many of its stations it is frequent to abundant; any decline which has occurred has been due to habitat destruction and, to some degree also, to collecting.

12/19 GB post 1960   1 NNR : 1 NCTR : 8 SSSI   TN = 6 R

Silene italica (L.) Pers.   Italian Catchfly

This probable introduction has persisted in a quarry in Kent for over 100 years. When first observed, it was abundant over a wider area; in 1973 the population comprised several hundred plants. It has also been recorded as a casual in Sussex, Surrey, Cambridgeshire, Monmouthshire and Midlothian.

1/1 GB post 1970   No conservation   TN = 11 V

Lychnis alpina L.   Alpine Catchfly

This perennial is native in Cumberland and Angus, but almost certainly introduced in its Hebridean station. In Cumberland, a relatively small population (17 plants in 1970) is restricted to two gullies which are difficult of access, while another population has not been seen since its original discovery there; in Angus over 200 plants were seen in 1972 but it is said to be less abundant than it was 20 years ago. In both vice-counties the populations have been reduced considerably by collecting by botanists and gardeners, and this depredation is thought to continue. The distribution of the species is limited to sites with a high concentration of heavy metals in the substratum. Formerly the species was also in Westmorland where it was last recorded in 1870.

3/4 GB post 1970   2 NNR : 1 SSSI   TN = 7 R

Lychnis viscaria L.   Sticky Catchfly

This perennial occurs on cliffs, dry rocks and rock debris, chiefly of igneous origin, in Radnorshire, Montgomeryshire, Kirkcudbrightshire, Selkirkshire, Roxburghshire, Midlothian, Stirlingshire and Perthshire. In Radnorshire, Midlothian, Perthshire and Roxburghshire the populations are vigorous and, in the last named, as yet unaffected by the conifer planting which has taken place, though the encroachment of *Ulex* will probably be detrimental eventually; in Montgomeryshire and Midlothian, quarrying has caused a decline though in the former, affected plants have been saved by transplanting; in Selkirkshire, a landslide destroyed much of the population leaving only eight plants. Now extinct in Fife and Angus.

12/19 GB post 1950   5 SSSI   TN = 6 R

Agrostemma githago L.   Corncockle

This annual arable weed was once widespread in cornfields throughout the British Isles. It was recorded from 104 out of 112 vice-counties in Great Britain and from 32 out of 40 in Ireland, as well as from the Channel Isles. In many nineteenth century local Floras it was reported as common and no localities were given. However the introduction of cleaner seed corn in the 1920s began a rapid decline. It was recorded from fewer than 200 10 kilometre squares during the period 1930-1960 which was the basis for the map in the *Atlas of the British Flora*. The species persisted in some areas until the 1950s but since 1960, perhaps as a result of the increased use of herbicides in agriculture, it has become extremely rare and has only been reported from Norfolk, Cambridgeshire and Moray as an arable weed. Other recent records are in or near gardens where it seems likely that it is an introduction with bird-seed.

4/ ∞ GB post 1960   No conservation   TN = 13 E
0/4   S post 1934
2/59   H post 1960

Dianthus gratianopolitanus

Dianthus gratianopolitanus Vill.   Cheddar Pink

This perennial is native only on, and near, Carboniferous limestone cliffs in Somerset. Plants are still fairly numerous but, where accessible, the species has diminished noticeably since the 1950s. In the past it was collected by gardeners and others, in some cases for sale, but apparently this practice has now stopped. It is also recorded as an introduction in about eight localities in Somerset, Oxfordshire, Gloucestershire, Worcestershire, Warwickshire and in Northern Ireland.

3/3 GB post 1960   2 SSSI   TN = 10 V
                        Scheduled species   R in Europe

Petrorhagia nanteuilii (Burnet) Ball & Heywood   Childling Pink
Kohlrauschia prolifera auct. brit. pro parte

Since 1950, this annual of sandy and gravelly places near the sea has been recorded as a native at two stations in Hampshire and two in Sussex, and is still common in Jersey; however, it has apparently become extinct in Hampshire since 1965. At one time it was thought that it also occurred as a native in the Isle of Wight, Kent, Middlesex, Berkshire, Buckinghamshire, and Norfolk, being casual at several other widespread stations, but in 1962 when *Petrorhagia nanteuilii* was first recognised in Britain, all these records were referred to *Petrorhagia prolifera* (L.) P.W. Ball & Heywood *sensu stricto*. In Sussex the plant, which is attractive, is reasonably safe as, though easily accessible, it finishes blooming early in the year before holiday-makers arrive.

1/4 GB post 1970   1 SSSI   TN = 13 E
1/1   S post 1960

Cerastium arcticum Lange   Shetland Mouse-ear
      subsp. **edmondstonii** (H. C. Watson) A. &. D. Love
C. nigrescens Edmondst.

This endemic subspecies is a perennial occurring on serpentine debris on only one island in Shetland. There is one colony which is rather scattered over parts of two 10 kilometre squares, though frequent in patches; in 1965, large numbers of plants were noted in flower. Agricultural reseeding in 1967 caused the destruction of a large section of the colony but the remainder is now protected in an NNR.

2/2 GB post 1970   1 NNR : 1 SSSI   TN = 7 R

Cerastium brachypetalum Pers.   Grey Mouse-ear

This presumably native annual species has been known since 1946 in a railway cutting in Bedfordshire where there are three colonies, one considerably larger than the others, but numbers fluctuate annually. A fourth colony has recently been found nearby, in Northamptonshire. As it will apparently survive as long as this main line is in use, no special conservation measures are necessary.

1/1 GB post 1970   No conservation   TN = 7 R

Holosteum umbellatum L.   Jagged Chickweed

This annual species of old walls, thatched roofs and banks was first recorded, around 1765, in Norfolk; it also occurred in Suffolk and Surrey and was last seen in the latter county around 1930, having become extinct in Suffolk and Norfolk in the nineteenth century. Extinction was almost certainly due to habitat destruction.

0/5 GB post 1930   **EXTINCT**

Sagina normaniana Lagerh.   Scottish Pearlwort

This perennial occurs on water-splashed rock ledges on mountains in Perthshire, Angus and Inverness-shire; formerly it was much more widespread in these counties, and was also recorded from Aberdeenshire, Argyllshire and the Isle of Skye. At present only seven populations are known but it may well be overlooked and under-recorded. Regarded by some authors as a hybrid, *S. procumbens* x *saginoides*.

6/18 GB post 1950   4 NNR : 1 SSSI   TN = 5 R

Sagina intermedia Fenzl   Snow Pearlwort

This dwarf perennial is now found only in the mountains of Perthshire and Aberdeenshire in four stations at one of which, in 1975, the population comprised only about four plants; formerly it also occurred in Banffshire. Its apparent decline may have been caused by collecting, but the species may be overlooked and under-recorded.

4/6 GB post 1950   1 NNR : 1 NT : 2 SSSI   TN = 6 R

Minuartia rubella (Wahlenb.) Hiern   Mountain Sandwort

Small populations of this perennial herb occur on base-rich rock ledges and detritus near the tops of a few mountains in Perthshire, Inverness-shire and Sutherland; it is almost certainly extinct in Shetland. At one site, the plants are distributed sparsely over 500 square metres at another the situation is precarious because of trampling by sheep and botanists.

5/6 GB post 1950   1 NT : 3 SSSI   TN = 6 R

Minuartia stricta

Minuartia stricta (Sw.) Hiern   Teesdale Sandwort

Confined to two calcareous flushes on one fell in Durham: several hundred plants occur in the main flush systems on the plateau, with only a few known outside it.

1/1 GB post 1970   1 NNR   TN = 9 V
                              Scheduled species

Arenaria norvegica Gunn. subsp. **norvegica**   Arctic Sandwort

This subspecies is a perennial of base-rich screes in Inverness-shire, Argyllshire, the Inner Hebrides, Sutherland and Shetland; whilst, overall, it is possibly rather sparsely distributed, fair-sized populations have been recorded in several places, for example in Inverness-shire and Sutherland. In Shetland most populations are large and reasonably safe. A single Irish population was recently discovered in Clare.

8/8 GB post 1960   3 NNR : 1 SSSI   TN = 6 R
1/1 H post 1960

Arenaria norvegica subsp. **anglica** Halliday   English Sandwort

This endemic winter annual or biennial subspecies is confined to tracks and depressions on limestone in one area in Yorkshire where it is present in two 10 kilometre squares. No protection is given and the plant is very vulnerable to agricultural activities.

2/2 GB post 1950   1 SSSI   TN = 10 E

Spergularia bocconii (Scheele) Aschers. & Graebn.

                              Boccone's Sand-spurrey

This annual or biennial herb is now only known from two localities in Cornwall where its situation is precarious. It was thought to have become extinct in 1959 and has only recently been refound. Formerly also recorded from Devon and the Isles of Scilly as a native, and from Suffolk and Glamorgan as an alien. It grows on dry sands, bare places, and in rock crevices near the coast; such habitats are subject to natural environmental change and this species may reappear in some of its other former localities.

In the Channel Isles it is still plentiful in Guernsey and Jersey, and may occur in Sark.

2/9 GB post 1970   No conservation   TN = 11 E
2/3   S post 1950

Polycarpon tetraphyllum (L.) L.   Four-leaved Allseed

A rare and local annual species of sandy and waste places; it has been recorded since 1950 from the Isles of Scilly, Cornwall, Devon and the Channel Isles, and previously also from Dorset. Although a very efficient coloniser of open ground, and thus common to abundant in some of its stations, it has nevertheless decreased because of habitat destruction. It also occurs occasionally as an introduction and has persisted for the last 30 years as a weed in one Surrey garden.

3/13 GB post 1950   1 SSSI   TN = 8 R
5/5   S post 1950

Corrigiola litoralis L.   Strapwort

This annual herb has only ever occurred as a native at one station in each of Devon and Cornwall, on shingly pool margins. In Devon it is locally abundant but fluctuates yearly; in Cornwall it is now extinct. It occurs elsewhere as an introduction, being constant and fairly abundant in such stations, particularly in Hertfordshire where it occurs on railway ballast. Its decrease and subsequent extinction in Cornwall was due to a lack of fluctuation in the water level. It has never been threatened by collectors because of its unattractive appearance.

1/2 GB post 1970   1 FSCR   TN = 9 V

Herniaria glabra L.   Smooth Rupturewort

An annual or biennial of dry sandy places, recorded as a native in 12 stations in Suffolk, Norfolk, Cambridgeshire and Lincolnshire; it was formerly also in Hampshire, Middlesex, Nottinghamshire and Cumberland and has occurred as a casual in several scattered localities. Persistent and sometimes abundant at six stations; elsewhere large populations have declined as the open habitat has become colonised with competing vegetation. Overall, the decline recorded has been due to habitat alteration or destruction.

8/15 GB post 1970   1 NNR : 5 SSSI   TN = 8 V

Herniaria ciliolata

Herniaria ciliolata Melderis   Fringed Rupturewort

An evergreen dwarf shrub of maritime sands and rocks in Cornwall and the Channel Isles; on the Lizard Peninsula of Cornwall it covers a very large area, being abundant wherever the soil is shallow in cliff-top grassland, regardless of geology, and extending inland in places; it actively colonises walls and is favoured by the formation of coastal footpaths; in Guernsey, it was locally plentiful in 1971. As it is very widespread, being tolerant of wide environmental extremes, for example waterlogging in winter, there is no conservation threat. The situation is, at least, stable and the plant may indeed be increasing. It is known as an introduction in Glamorgan.

2/4 GB post 1970   1 NNR : 3 SSSI   TN = 8 R
3/3   S post 1950

Scleranthus perennis L. subsp. **perennis**   Perennial Knawel

This subspecies is confined to one rocky locality in Radnorshire where its population fluctuates and is often reduced to one plant. It is constantly threatened by collectors and by visitor pressure.

1/1 GB post 1970   1 SSSI   TN = 10 E

Scleranthus perennis subsp. **prostratus** P.D.Sell   Perennial Knawel

This endemic subspecies is a very local plant of sandy heaths in the Breckland of Suffolk and Norfolk. It now occurs in only four stations, in the majority of which it is threatened by agriculture or housing development.

2/11 GB post 1970   3 SSSI   TN = 11 E

Chenopodium vulvaria L.   Stinking Goosefoot

This mealy annual was formerly widespread in south and east England as a native or persistent introduction and was occasionally recorded as a casual elsewhere. It was particularly associated with waste ground near the sea and may have been a natural component of strand-line vegetation. Since 1960 it has been recorded from only about 15 10 kilometre squares in England, from rubbish tips and similar habitats, except for a few localities along the south coast, in Dorset, Sussex and Kent. Long known and still persistent in Guernsey and Jersey in the Channel Isles.

15/95 GB post 1960   No conservation   TN = 9 V
  2/3   S post 1970

Atriplex longipes Drejer

This procumbent annual species of salt-marshes has only recently been recognised in Britain and is now known from two or three localities in Somerset and Norfolk.

2/3 GB post 1970   No conservation   TN = 9R

Atriplex praecox Hülphers

This procumbent annual of fine shingle beaches has only recently been recognised as a native British species. Since 1975 it has been recorded from three localities in Ross, Sutherland and Shetland and may well be more widespread.

3/3 GB post 1970   No conservation   TN = 8 R

Halimione pedunculata (L.) Aellen

This species of muddy salt-marshes was first recorded in Britain in 1650 and occurred in at least 16 localities in Kent, Suffolk, Norfolk, Cambridge-shire and Lincolnshire, but has been extinct since about 1935. On the Continent, it is rare and decreasing. Two casual occurrences are known — in Glamorgan and Durham.

0/14 GB post 1935   **EXTINCT**

Lavatera cretica L.   Smaller Tree-mallow

This annual or biennial herb of waste ground, old quarries, roadsides and bulbfields is regularly recorded only from the Isles of Scilly and the Channel Isles. There are also records from Cornwall where, however, it has not been seen since early this century. It has also occasionally been recorded as a casual. In the Isles of Scilly there are considerable changes in the size of the population from year to year largely related to seasonal climatic conditions.

1/3 GB post 1970   1 SSSI   TN = 7 R
2/2 S post 1950

Althaea hirsuta L.   Rough Marsh-mallow

An annual or biennial herb of field borders, scrub and wood margins, believed to be native only in Somerset, where it was last seen in 1955, and in Kent, where two localities are known. It is widespread as a casual but has had a chequered existence for the past 150 years; various factors, such as unsuitable crops and poor weather at germination time may have been responsible for this.

1/2 GB post 1970   1 SSSI   TN = 12 E

Geranium purpureum

Geranium purpureum Vill. subsp. **purpureum**   Little-Robin

This upright annual or biennial herb is found in rocky and stony places, and in open hedgebanks, usually near the sea, and always in a rich soil (normally derived from limestone). Recently only recorded from Cornwall, Dorset, Hampshire, Gloucestershire, the Channel Isles and Cork but previously known also in Devon, Somerset, Sussex and Carmarthen, and from Waterford in Ireland. This apparent decline may not be the true situation: the taxon is almost certainly under-recorded.

8/22 GB post 1950	No conservation	TN = 10 V
3/3 S post 1970		
1/2 H post 1970		

Geranium purpureum subsp. **forsteri** (Wilmott) H.G. Bak.

This endemic procumbent subspecies grows only in the stabilised area at the rear of shingle beaches in Hampshire and Sussex. Not seen in Guernsey in the Channel Isles since 1914 and apparently declining elsewhere, and is an example of outlying populations approaching extinction.

3/7 GB post 1950	No conservation	TN = 11 V
0/1 S post 1914		

Buxus sempervirens L.   Box

Over 20 localities of this evergreen shrub are known in beechwoods and scrub on chalk and oolitic limestone in Kent, Surrey, Berkshire, Buckinghamshire and Gloucestershire, in all of which the species is locally abundant and almost certainly native, though it is widely planted and often naturalised elsewhere. In at least one station, regeneration is endangered by trampling, but generally there is no sign of decline in the acknowledged native stations.

7/8 GB post 1960	3 NT : 9 SSSI	TN = 7 R

Genista pilosa L.   Hairy Greenweed

This creeping perennial occurs on cliffs and dry sandy and gravelly heaths on poor soils in over 40 localities in Cornwall, Sussex, Pembrokeshire and Merionethshire, but is now extinct in Kent and Suffolk. In Cornwall it is generally abundant; in Pembrokeshire there are many small but several quite extensive populations; there are scattered plants only in Merionethshire. The species has suffered from oil pollution in Cornwall and from burning in Sussex and Pembrokeshire but is reasonably safe in its coastal localities.

11/19 GB post 1960	1 NNR : 1 NCTR : 3 NT : 14 SSSI	TN = 5 R

Ononis reclinata L.   Small Restharrow

This annual maritime species occurs on limestone in Devon and Glamorgan where it is rare, in Pembrokeshire where one of the two colonies known may be extinct, and on two of the Channel Isles; it has recently been refound in Wigtownshire where it had not been seen since 1835.

5/7 GB post 1950   1 NNR : 2 SSSI   TN = 8 V
2/3   S post 1950

Trifolium stellatum L.   Starry Clover

This annual, which appears to have been recorded first in Britain around 1700, seems to have persisted for a period of over 170 years in one locality in Sussex, where it still occurs in small quantities in suburban surroundings. The site is well known and subject to exploitation. As a casual, it has occurred elsewhere in Sussex and also in Kent, Essex, Suffolk, Glamorgan, the Channel Isles (Alderney) and in Down, Northern Ireland.

1/1 GB post 1960   No conservation   TN = 12 E

Trifolium molinerii Balb.   Long-headed Clover

This annual is native in seven stations in Cornwall, where it is locally abundant over several hundred square metres of grassland on cliff-tops exposed to sea spray, and in Jersey in the Channel Isles; it has been recorded as a casual elsewhere in Cornwall, and also in Devon, Dorset, Hampshire and Suffolk, all before 1900.

2/3 GB post 1960   1 NNR   TN = 8 R
1/1   S post 1960

Trifolium bocconei Savi   Twin-flowered Clover

This annual occurs in grassy places in five localities in Cornwall, and in Jersey in the Channel Isles; it is widely scattered and destruction of the habitat may be the only threat to its survival.

2/3 GB post 1960   1 NNR : 2SSSI   TN = 9 R
1/1   S post 1960

Trifolium strictum L.   Upright Clover

Since 1960, this annual of grassy places has been recorded from six stations in Cornwall and one in Jersey in the Channel Isles: it now appears to be extinct in Guernsey. It is very locally abundant in Cornwall and benefits from burning, grazing and human activity, thus some population pressure is desirable. There are also records from Radnorshire (1936) and Anglesey (1840) though there are doubts about the status of this species in both these stations. It has also occurred definitely as a casual in Sussex and Midlothian.

2/4 GB post 1960   1 NNR   TN = 8 R
1/2   S post 1960

Lotus angustissimus

Lotus angustissimus L.   Slender Bird's-foot-trefoil

This perennial occurs as a native in about 16 localities in dry maritime grassland in Cornwall, Devon, Hampshire and Kent, also in all the larger Channel Isles; formerly it was much more abundant, particularly in Cornwall and Devon, and was also recorded from Sussex and Caernarvonshire. It has occurred as an introduction in Surrey, Norfolk, and Down in Northern Ireland.

13/46 GB post 1950   1 NT : 5 SSSI   TN = 9 R
 5/5   S post 1950

Tetragonolobus maritimus (L.) Roth   Dragon's-teeth

This alien perennial is well naturalised in about nine localities in Kent, Essex, Berkshire, Buckinghamshire and Gloucestershire but is now extinct in Hampshire; it has been recorded as a casual also in Kent and Gloucestershire, and in Dorset, Oxfordshire, Staffordshire, Glamorgan and Lincolnshire. So far as is known, no threat exists except perhaps to the habitat — rough calcareous grassland.

9/9 GB post 1950   1 SSSI   TN = 9 R

Astragalus alpinus L.   Alpine Milk-vetch

A rare mountain perennial herb of which four colonies are known in Perthshire, Angus and Aberdeenshire. The sizes of the colonies vary considerably; in one thousands of plants occur in an area of about 600 square metres, but in another the patch is only one square metre. Although rather accessible in one station, the species is apparently adequately protected. It suffers severely from grazing and trampling and was possibly over-collected in the past.

4/4 GB post 1960   1 NNR : 3 SSSI   TN = 8 R

Oxytropis halleri Bunge   Purple Oxytropis

About 21 populations of this perennial are known to occur on dry rocky cliff pastures in Perthshire, Argyllshire, Ross-shire and Sutherland; the species was formerly also recorded from Wigtownshire, Fife, Angus and Caithness. Populations range in size from very small — about 10 plants — to 100,000 or more in some Sutherland localities. The species is subject to heavy grazing in some of its stations, though the total effect of this is unknown.

11/17 GB post 1960   1 NNR : 3 SSSI   TN = 6 R

Oxytropis campestris (L.) DC.   Yellow Oxytropis

This perennial grows on rock ledges in perhaps up to six localities in

Perthshire, Angus and Kintyre. All populations are large, and one, in Angus, is known to have increased. In Kintyre, feral goats instead of being destructive, as presumed, may indeed help by keeping other vegetation down; here the species was originally recorded in error as *O. halleri* Bunge.
3/3 GB post 1970   1 NNR : 1 SSSI   TN = 6 R

Ornithopus pinnatus (Mill.) Druce   Orange Bird's-foot

This annual is now confined to the Isles of Scilly where many stations have been recorded and where it occurs on all but one of the larger islands, and to the Channel Isles, where, however, it has become extinct this century on two islands. Also recorded from Cornwall and as a casual from Glamorgan. It grows in short open turf on sandy or granite soil and is probably able to take advantage of any opportunity to spread.
1/2 GB post 1970   1 SSSI   TN = 6 R
3/5   S post 1970

Rubus arcticus L.

This perennial species, now apparently extinct, may have occurred as a native in three or four localities in the Scottish Highlands but has not been seen since the beginning of the nineteenth century. The reason for its disappearance is not known.
0/2 GB post 1800   **EXTINCT**

Potentilla fruticosa L.   Shrubby Cinquefoil

About 12 populations of this perennial shrub have been recorded on damp rocky ground, nearly always basic, in Yorkshire, Durham, Westmorland and Cumberland; in Ireland it is found in Clare and Mayo but has not been recorded recently from Galway. It is abundant in some stations whilst, in others, only small colonies are known; it has definitely suffered from collecting in both England and Ireland.
6/7 GB post 1960   6 NNR : 1 SSSI   TN = 6 R
5/8   H post 1960

Potentilla rupestris L.   Rock Cinquefoil

Only three populations of this perennial herb are known, all on basic rocks. In Radnorshire the species is abundant in one station and is seeding well; in Montgomeryshire it is just surviving though threatened by quarrying; in Sutherland there is a colony of considerable size thickly scattered over the cliffs, though this is now said to be endangered. It was so heavily collected by gardeners and botanists in Radnorshire in the past that it was once thought to be extinct until a few plants were found around 1940.
3/3 GB post 1970   1 SSSI   TN = 9 V

Alchemilla glaucescens

Alchemilla glaucescens Wallr.

This local species of limestone grassland is recorded from Yorkshire, Ross, Sutherland and from Leitrim in Ireland. It occurs altogether in at least 13 localities. It has also been recorded as a garden escape in Surrey, S.W. Yorkshire and Midlothian, and in error elsewhere.

| 9/15 GB post 1950 | 1 NNR : 3 SSSI | TN = 7 R |
| 1/1   H post 1950 | | |

Alchemilla subcrenata Buser

This very rare northern montane meadow species was discovered in Britain in 1951 and is confined to Teesdale (both Yorkshire and Durham) where only five localities are known at this, its most westerly limit in Europe.

| 2/2 GB post 1950 | 1 SSSI | TN = 9 V |

Alchemilla minima Walters

This, the only endemic British species of the genus, is known definitely from closely grazed *Festuca-Agrostis* grassland on two mountains in the Craven district of the north Pennines in Yorkshire. Other records from this area await confirmation.

| 2/2 GB post 1950 | 2 SSSI | TN = 9 V |

Alchemilla monticola Opiz

This northern montane species is frequent to abundant on roadsides and in hay-meadows in Yorkshire and Durham (in both Weardale and Teesdale) where it is certainly native in all its known stations. It has also been recorded as a casual in Surrey and Buckinghamshire though it has now apparently become extinct in the former. Other records have been made in error.

| 9/9 GB post 1950 | 1 NNR : 2 SSSI | TN = 5 R |

Alchemilla acutiloba Opiz

This northern montane species is confined to the Teesdale/Weardale area of Durham where it is occasional on roadsides and in hay-meadows; over 20 populations have been recorded.

| 11/11 GB post 1950 | No conservation | TN = 6 R |

Cotoneaster integerrimus Medic.   Wild Cotoneaster

Only about five plants of this deciduous shrub are known, forming one very dispersed colony on limestone rocks in Caernarvonshire where, around 1900, the species had become virtually extinct through over-collecting and grazing by goats and sheep.

1/1 GB post 1970   1 SSSI   TN = 11 E

Sorbus pseudofennica E.F.Warb.

A small endemic tree found only on a steep granite stream bank in Glen Catacol, Arran, where few plants are known. It is closely related to the Fennoscandian species *S. hybrida* L.

1/1 GB post 1970   1 NNR   TN = 6 R

Sorbus arranensis Hedl.

This very rare endemic tree is recorded from the steep granite stream banks of two glens in Arran, in one of which only a few plants are reported.

1/1 GB post 1970   1 NNR   TN = 6 R

Sorbus leyana Wilmott.

This very rare endemic shrub is confined to two Carboniferous limestone crags in Brecon: the total world population does not appear to exceed ten trees.

2/2 GB post 1970   1 NNR : 1 FNR   TN = 6 R

Sorbus minima (A.Ley) Hedl.

This rare endemic shrub is confined to Breconshire where over 300 trees have been recorded within a very small area, all on Carboniferous limestone crags.

1/1 GB post 1970   1 NNR   TN = 5 R

Sorbus anglica Hedl.

This endemic shrub has been recorded from a number of localities in Devon, Somerset, Gloucestershire, Monmouthshire, Herefordshire, Shropshire, Brecon, Montgomeryshire and Denbighshire, and, in Ireland, from Killarney. It occurs in scrub and woodland mainly on Carboniferous limestone slopes and cliffs, and varies from one population to another.

12/12 GB post 1950   3 NNR : 1 FNR : 7 SSSI   TN = 3 R
1/1   H post 1950

Sorbus leptophylla E.F.Warb.

This rare and local shrub is known with certainty only on wooded Carboniferous limestone cliffs in Brecon where it replaces *S. aria* (L.) Crantz; it may also occur in Montgomeryshire, but this record still requires confirmation.

2/2 GB post 1970   1 NNR : 1 SSSI   TN = 6 R

Sorbus wilmottiana

Sorbus wilmottiana E.F.Warb.

This endemic tree is confined to rocky limestone scrub and woodland on both sides of the Avon Gorge in Somerset and Gloucestershire.

1/1 GB post 1970	1 NNR	TN = 7 R

Sorbus eminens E.F.Warb.

This rare endemic tree occurs in woods on Carboniferous limestone in Somerset, Gloucestershire and Herefordshire and is possibly also still in Monmouthshire where, however, it has not been recorded since the beginning of the century.

2/3 GB post 1970	1 NNR : 1 SSSI	TN = 7 R

Sorbus porrigentiformis E.F.Warb.

This probably endemic shrub occurs on crags and in rocky woods on limestone in about 20 localities in Devon, Somerset, Gloucestershire, Monmouthshire, Herefordshire, Glamorgan and Brecon. Formerly recorded from Radnorshire, Carmarthenshire and Caernarvonshire where it may well still grow.

14/19 GB post 1950	3 NNR	TN = 4 R

Sorbus lancastriensis E.F.Warb.

This endemic shrub has been recorded since 1950 in about six localities in Lancashire and Westmorland and is possibly still in four others where it has not been confirmed recently: it occurs in woodland on slopes and cliffs of Carboniferous limestone.

3/4 GB post 1950	2 SSSI	TN = 6 R

Sorbus vexans E.F.Warb.

This rare endemic tree has been recorded from a small stretch, about 10 km long, of the coasts of Devon and Somerset where it is found in oak woods: unlike the majority of other rare endemic species, it does not grow on limestone.

2/2 GB post 1950	1 SSSI	TN = 10 V

Sorbus bristoliensis Wilmott

This endemic tree occurs in rocky woods and scrub on Carboniferous limestone in the Avon Gorge in both Somerset and Gloucestershire, where over 100 plants are known.

1/1 GB post 1970	1 NNR	TN = 6 R

Sorbus subcuneata Wilmott

This small graceful endemic tree is confined to sessile oakwoods over a small area of the coasts of Devon and Somerset. In 1973, two plants were seen, though more may exist, in one well-known locality but it is now apparently extinct at another following felling and a fire. Unlike the majority of other rare endemic species, it does not grow on limestone.

2/3 GB post 1970 1 SSSI TN = 10 V

Pyrus cordata Desv.

A hedgerow shrub now confined to two stations in Devon; it was formerly also in two localities in Cornwall where it is now extinct, though the reason for this is unknown. Because of threatened industrial development it has been necessary to move the plant from one locality to safety.

2/4 GB post 1960 No conservation TN = 13 E

Crassula aquatica (L.) Schönl.

This species, which is possibly native, occurs on the muddy margins of pools. It was first recorded in Yorkshire in 1921, but had disappeared by 1945, thus apparently becoming extinct in the British Isles. It was never refound in its original station, but in 1969 it was recorded for the first time in Inverness-shire, in small quantity over an area of six square metres, suggesting that it may have been there for some time.

1/2 GB post 1970 No conservation TN = 9 V

Saxifraga hirculus L.   Marsh Saxifrage

A perennial herb of wet base-rich hollows in moorland and bogs. Now confined to about 20 localities in Yorkshire, Westmorland, Cumberland, Midlothian, Aberdeenshire, Banff and Inverness-shire, and to Mayo and Antrim in Ireland. Formerly it occurred in Cheshire, Durham, Lanark, Peebles, Berwickshire, Perthshire, Kincardineshire and Caithness, and in several other Irish counties. A species which has declined abruptly due to drainage and the improvement of upland pastures. In Britain some colonies are very small, others large, but numbers fluctuate considerably from year to year at certain stations.

10/21 GB post 1950 3 NNR TN = 5 R
 2/11 H post 1950

Saxifraga cernua

Saxifraga cernua L.   Drooping Saxifrage

A high-montane species occurring in corries and crevices in basic rock, in three main stations — with at least four colonies — in Perthshire, Inverness-shire and Argyllshire, in all of which populations are less than 100. Formerly there were possibly two other stations in Perthshire where it is almost certainly extinct. Despite its scarcity, it is still subject to depredation (as it has been, by botanists and alpine gardeners, for over a century) and there is evidence of collecting in at least one population during the last ten years.

3/5 GB post 1970   4 SSSI   TN = 8 V
                           Scheduled species

Saxifraga rivularis L.   Highland Saxifrage

This perennial herb of wet rocks on high mountains is known from 17 localities in Perthshire, Aberdeenshire, Banffshire, Inverness-shire, Argyllshire and Ross. In several of these 100 or more plants have been seen but in others as few as three have been reported. Previously more widespread and recorded from Angus, but not seen there for nearly 150 years. Many of its remaining sites are however protected by their inaccessibility except to skilled mountaineers.

12/16 GB post 1960   6 NNR : 4 SSSI   TN = 5 R

Saxifraga cespitosa L.   Tufted Saxifrage

A mountain species known from ten stations in Caernarvonshire, Aberdeenshire, Banffshire, Inverness-shire and Ross mainly at over 3000 ft. (920 m.). It is one of the rarest of all British mountain saxifrages always occurring sparsely, in small quantities, and is safe in certain places only because it is inaccessible. Collecting has certainly caused a decrease within colonies in the past, whilst the drought of the summer of 1976 almost destroyed the North Wales population.

9/9 GB post 1950   2 NNR : 3 SSSI   TN = 6 R
                         Scheduled species

Saxifraga rosacea Moench   Irish Saxifrage

This alpine species was always very rare in its sole British station in Caernarvonshire where it was first discovered in 1821, and it became extinct about 1900. It is still abundant in various places in Ireland on the mountains and sea cliffs of Kerry, Clare, Galway and possibly Mayo, and it is also found on mountains on the border of Tipperary and Limerick.

0/1   GB post 1900   **EXTINCT**
18/19   H post 1930

Lythrum hyssopifolia L.   Grass-poly

This annual of bare ground flooded in winter is very irregular in its occurrence, and may not have declined as much as the figures suggest. Though formerly widespread throughout S.E. England the only recent records are from Essex, Cambridgeshire, Bedfordshire and Huntingdonshire, and Jersey in the Channel Isles. It only appears regularly in Cambridgeshire and in Jersey. Drainage and modern agricultural methods are reducing the number of suitable habitats.

3/38 GB post 1950   No conservation   TN = 11E
1/2   S post 1950

Ludwigia palustris (L.) Ell.   Hampshire-purslane

This rare annual or perennial aquatic herb is confined to about 12 localities in shallow pools in the New Forest in Hampshire and to a single locality recently discovered in Epping Forest in Essex. Formerly it also occurred in Sussex, but not seen in that county for about 100 years, and in Jersey in the Channel Isles where it was last recorded in 1926. Established as an introduction in a canal in Lancashire. Most of the decline has been caused by the infilling and over-growth of the shallow ponds and streams in which it occurred, but it is often still abundant and sometimes dominant in its remaining localities.

5/9 GB post 1950   2 SSSI   TN = 7 R
0/1   S post 1926

Oenothera stricta Ledeb.   Fragrant Evening-primrose

This annual or biennial herb was introduced from South America but is now well established in a few sand dune areas in the southern half of Britain and in the Channel Isles. It is also recorded inland as a casual on rubbish-tips and similar places.

10/29 GB post 1960   1 NNR : 1 NCRT : 2 SSSI   TN = 9 R
 4/4   S post 1970

Eryngium campestre L.   Field Eryngo

A perennial of dry grassy places usually on the coast which was native or long established in Cornwall, Devon, Somerset, Wiltshire, Hampshire, Sussex, Kent, Suffolk and the Channel Isles. It has also been widely recorded as an obvious introduction. Now restricted to Cornwall, Devon, Hampshire and Guernsey, though it has recently been successfully re-introduced into its Wiltshire locality. The two Devon populations are both large and well protected and said to be increasing.

4/15 GB post 1950   1 SSSI   TN = 11 V
1/1   S post 1960

Caucalis platycarpos

Caucalis platycarpos L. Small Bur-parsley

This erect annual formerly occurred as a casual or naturalised alien through-out much of lowland England, especially on chalky soils. During the period 1930-1960 it was only recorded from 11 10 kilometre squares: there have been no records since 1962 and it is possible that this species is now extinct.

3/81 GB post 1960 No conservation TN = 12 E

Physospermum cornubiense (L.) DC. Bladderseed

A locally abundant perennial in cornfields, hedgebanks, scrub and woods in Cornwall, Devon and Buckinghamshire which is not apparently threatened.

7/11 GB post 1950 1 SSSI TN = 7 R

Bupleurum rotundifolium L. Thorow-wax

This annual cornfield weed was formerly widespread in England south and east of a line joining the mouths of the River Severn and the River Tees, and was reported from 53 vice-counties. However, the introduction of cleaner seed corn in the 1920s began a rapid decline. It was recorded from only 17 10 kilometre squares during the period 1930-1960 which was the basis of the map in the *Atlas of the British Flora*. Since 1960 the only records are from Devon, Wiltshire, Norfolk and Rutland and it seems possible that this species is now extinct in Britain. Records are frequently received but they invariably turn out to be the closely related *B. intermedium* Poiret (*B. lancifolium* auct., non Hornem.) which is commonly introduced with wild bird seed and is found in gardens or on refuse tips.

8/150+ GB post 1960 No conservation TN = 10 E

Bupleurum baldense Turra Small Hare's-ear

A species of dry banks, rocky slopes and grey dunes near the sea. Formerly known from four localities in England it is now confined to two, one in Devon and another in Sussex. It also occurs in all the major Channel Isles except Sark, and has been recorded as a casual from Surrey, Leicester-shire and East Yorkshire.

2/3 GB post 1960 1 LNR : 1 SSSI TN = 10 V
4/4 S post 1960

Bupleurum falcatum L. Sickle-leaved Hare's-ear

When first found in Essex in 1831 this perennial was scattered over several miles between Ongar and Chelmsford. By the middle of this century the population had been reduced to a short stretch of damp verge and ditch bank and in 1962 this site was destroyed by hedgerow clearance and ditch cleaning. Seed from this site has now successfully established a population in an Essex Naturalists' Trust Reserve. Also reported from Surrey in the middle of the nineteenth century but it lasted only a few years and has not been recorded recently.

0/2 GB post 1962   **EXTINCT**

Trinia glauca (L.) Dumort.   Honewort

This perennial is fairly abundant in dry limestone grassland in all 10 localities from which it is known in Devon, Somerset and Gloucestershire. There has been little apparent decline and this species is sufficiently well protected to be safe from losses caused by habitat destruction.

5/5 GB post 1960   1 NNR : 1 LNR : 5 SSSI   TN = 5 R

Apium repens (Jacq.) Lag.   Creeping Marshwort

This creeping perennial is now represented in Britain by four populations in Oxfordshire where it is frequent, growing in old meadows, ditches and shallow ponds. All these populations are threatened by drainage and by hybridisation with *A. nodiflorum* (L.) Lag. Formerly also in one locality in South-east Yorkshire. Elsewhere, in Suffolk, Norfolk, East Yorkshire, the Lothians, Fife and Kintyre it is only the hybrid which now occurs.

1/2 GB post 1960   1 SSSI   TN = 10 E
                                                            V in Europe

Petroselinum segetum (L.) Koch   Corn Parsley

This slender biennial occurs in arable fields, in hedgebanks and on road-sides almost entirely south and east of a line joining the Humber to the River Severn. Whilst there is some evidence of decline locally it is still widespread and in no way threatened. It is however under threat in Europe as a whole and is included in the IUCN list.

                                        IUCN List, V in Europe

Bunium bulbocastanum L.   Great Pignut

This perennial of rough calcareous grassland, arable and banks is confined to a compact area on the borders of Hertfordshire, Buckinghamshire and Bedfordshire, and there are two outliers in Cambridgeshire. Within the main area it is known from at least 30 localities and no significant change in frequency has been noted during the last 30 years. Also once occurred for a short while as a cornfield weed in Gloucestershire.

11/15 GB post 1960   2 SSSI   TN = 5 R

Seseli libanotis (L.) Koch   Moon Carrot

This perennial occurs in rough grassy and bushy places on chalk hills in Sussex, Hertfordshire, Cambridgeshire and Bedfordshire: seven stations are known in most of which it is abundant but the Hertfordshire locality now has only a single plant.

4/6 GB post 1960   1 NNR : 1 NCTR : 2 SSSI   TN = 8 R

Selinum carvifolia (L.) L.   Cambridge Milk-parsley

This perennial of fens and damp meadows is found now in only two localities in Cambridgeshire, in both of which it is abundant. Formerly also recorded from Lincolnshire and Nottinghamshire but both sites have been destroyed by drainage since 1930.

2/5 GB post 1970   1 NNR : 1 SSSI   TN = 9 V

Peucedanum officinale L.   Hog's Fennel

This perennial occurs on the banks of creeks near the sea in Kent and Essex only. About 15 populations are known, some of which are very large and form a more or less continuous stand. It has been extinct in Sussex since 1695 soon after it was first recorded there.

5/6 GB post 1950   NCTR   TN = 6 R

Euphorbia peplis L.   Purple Spurge

An annual of sandy and shingly beaches which formerly occurred in several places in Cornwall, Isles of Scilly, Devon, Somerset, Dorset, Isle of Wight, Kent, Cardiganshire and in one locality in Ireland. It is probably now extinct in all of these except on Lundy Island off the coast of Devon. It has occurred recently on two of the Channel Isles but can only be seen regularly in Alderney. Its extinction in at least some, if not all, of these stations was due to the destruction of the habitat by natural causes. Trampling by bathers has seriously affected the survival of the species on some beaches.

1/22 GB post 1960   No conservation   TN = 9 E
2/5   S post 1960
0/1   H post 1830

Euphorbia villosa Waldst. & Kit.   Hairy Spurge

E. pilosa auct. eur., non L.

This perennial was first recorded from its sole station, a wood in Somerset, around 1576, and it persisted there until about 1924. It was plentiful in the 1880s but the population size fluctuated with coppicing, being greatest just after clearance; then followed an annual decrease as the brushwood grew up, until little or none might be found in certain seasons. Cessation of coppicing about 1900 finally led to its extinction.

0/1 GB post 1924   **EXTINCT**

Euphorbia hyberna L.   Irish Spurge

A perennial of shady woods, and stream banks on lime-free soils near the sea in Cornwall, Devon and Somerset. In Ireland it grows in both coastal and inland stations. It is frequent to abundant in two of its three English stations, but is somewhat vulnerable unless woodland management is practised.

2/2   GB post 1960   2 SSSI   TN = 8 R
86/103   H post 1930

Euphorbia serrulata Thuill.   Upright Spurge

E. stricta L.

This annual occurs in woodland clearings on limestone in Gloucestershire and Monmouthshire. About ten colonies are known. In Gloucestershire it is locally abundant and appears to be spreading, whilst in Monmouthshire it is rare in three stations but abundant in the fourth where coppicing has recently been carried out. It is also recorded as an established introduction in Somerset and as a rare casual in Sussex, Worcestershire and Leicester-shire.

5/5 GB post 1960   1 FNR : 1 SSSI   TN = 7 R

Polygonum maritimum L.   Sea Knotgrass

This procumbent perennial occurs on sand and fine shingle at and above high-tide level. Since 1966 it has been recorded at two stations in Cornwall, having previously grown as a native in the Isles of Scilly, Devon, Somerset and Hampshire, and as an introduction in Glamorgan. It is still to be found sparingly on Herm in the Channel Isles, and was recently found for the first time in Ireland in Waterford. These fluctuations in distribution are characteristic of strand-line species.

2/11 GB post 1960   No conservation   TN = 9 E
1/3   S post 1970
1/1   H post 1970

Koenigia islandica L.   Iceland-purslane

This annual occurs on two Hebridean islands on bare stony ground amongst open vegetation at 1500-2360 ft. (450-725 m.). Though not recognised as occurring in this country until 1950 many locations have now been discovered. It is locally abundant and in no danger.

5/5 GB post 1960   3 SSSI   TN = 3 R

Rumex aquaticus

Rumex aquaticus L.   Scottish Dock

Over 20 colonies of this perennial are known at the margins of alder swamps around Loch Lomond in Stirlingshire and Dunbartonshire. It occurs in fair quantity and hybridizes with *Rumex obtusifolius* L. There is no apparent threat.

3/3 GB post 1970   1 NNR   TN = 6 R

Rumex rupestris Le Gall   Shore Dock

About 12 stations of this perennial are known on sea cliffs, rocky shores and in dune slacks in Cornwall, Isles of Scilly, Devon, Dorset, Glamorgan, Pembrokeshire, Anglesey and the Channel Isles. Though frequent in Scilly, overall it is not very abundant and is somewhat sporadic in its appearances. It is vulnerable to coastal recreation pressures.

11/27 GB post 1950   3 NNR : 2 SSSI   TN = 8 V
  3/4   S post 1970   V in Europe

Salix lanata L.   Woolly Willow

This shrub occurs on damp ledges of basic rocks on mountains in 14 localities in Perthshire, Angus, Aberdeenshire and Inverness-shire. It is frequent to abundant in most of these with many young plants present as well as many established mature ones. Apparently no real threat exists and it has some degree of protection because of difficulties of access.

10/11 GB post 1950   3 NNR : 5 SSSI   TN = 4 R

Ledum groenlandicum Oeder   Labrador-tea

This shrub is possibly native in bogs in one area of Perthshire but it also occurs as a rare escape from cultivation in seven localities in Surrey, Derbyshire, Lancashire, Yorkshire, Cumberland and Kirkcudbrightshire. Most populations are small though long established. In Surrey it is now limited to a small area though it was said to be abundant when first discovered.

9/10 GB post 1960   1 NNR   TN = 6 R

Phyllodoce caerulea (L.) Bab.   Blue Heath

This high-montane rocky moorland evergreen shrub is known from only about three localities in Perthshire and Inverness-shire. It was first recorded in Perthshire where the single large patch is unusually precarious having been collected by botanists over the past 150 years, and by gardeners for propagation and sale. In Inverness-shire, both colonies are very small. It is partly protected by the resemblance of its leaves to those of Crowberry, *Empetrum nigrum* L., and partly by its shy flowering during a short season.

3/3 GB post 1960   2 SSSI   TN = 9 V
Scheduled species

Erica ciliaris L.   Dorset Heath

This small shrub is abundant on heaths in Cornwall, Devon and Dorset where altogether over 40 localities are still known of which approximately two thirds are in Cornwall. It has also been found recently in a single locality in Galway, Western Ireland, where it covers only a few square metres. However the species has declined sharply in area mainly as a result of habitat destruction, especially in Dorset, but partly as a result of natural hybridization with *Erica tetralix* L.

9/17 GB post 1950   2 NNR : 1 NCTR : 1 SSSI   TN = 8 R
1/1   H post 1960

Erica vagans L.   Cornish Heath

This low shrub occurs in about 10 stations on heaths in Cornwall where it is often locally dominant. There is also a single locality in Fermanagh in Ireland where about 500 plants have been recorded in what appears to be a native situation. Not refound in Waterford where its status was uncertain. Threatened by the extension of farming, building of holiday camps, the development of satellite-tracking stations and afforestation.

6/8 GB post 1960   1 NNR : 2 SSSI   TN = 8 R
1/2   H post 1970

Moneses uniflora (L.) A.Gray   One-flowered Wintergreen

This evergreen perennial herb is confined to pinewoods in Banffshire, Moray, Inverness-shire, Ross and Sutherland: the exact number of extant localities is not known but at least 17 are recorded. Formerly also reported from Perthshire, Aberdeenshire, the Inner and Outer Hebrides. Its disappearance from several stations has been due to the effects of felling and other forestry operations.

13/24 GB post 1950   2 SSSI   TN = 7 R

Diapensia lapponica L.   Diapensia

This prostrate shrub has been recorded from only one exposed mountaintop in Inverness-shire, where it was first found in 1951. It occurs in moderate quantity over a very small area; it is reported to be overcollected, and also much damaged by deer.

1/1 GB post 1970   No conservation   TN = 10 V
Scheduled species

Limonium bellidifolium

Limonium bellidifolium (Gouan) Dumort.   Matted Sea-lavender

This perennial herb occurs locally in the drier parts of sandy salt-marshes in Norfolk and Lincolnshire: it has been extinct in Cambridgeshire since the nineteenth century. In Norfolk it is characteristic of the sandy margins of salt marsh westward from Blakeney to Wolferton, and much more plentiful than in Lincolnshire where only two small patches are known, one of which has diminished since 1952 when it was first found.

6/10 GB post 1960   2 NNR : 1 NCTR : 3 NT   TN = 6 R

Limonium recurvum C.E. Salmon

This endemic perennial is known from only a single locality on maritime rocks and cliffs in Dorset, from which thousands of plants were recently reported. This species is protected from collectors in part by its resemblance to *L. binervosum* (G.E.Sm.) C.E. Salmon to which it is closely related, but there is growing public pressure from a nearby camp site.

1/1 GB post 1970   1 SSSI   TN = 10 E

Limonium transwallianum (Pugsl.) Pugsl.

This endemic to the British Isles is now probably confined in Britain to maritime cliffs in Pembrokeshire where two localities are known, from one of which hundreds of plants are reported. A record from the coast of Devon has not been recently confirmed. Also known from two stations in West Ireland, in the Burren and on the Arran Isles. One of the Welsh sites is protected by being within Ministry of Defence land.

2/3 GB post 1970   1 SSSI   TN = 9 R
2/3   H post 1970   R in Europe

Limonium paradoxum Pugsl.

This endemic to the British Isles is confined in Britain to one locality on basic igneous rocks on the Pembrokeshire coast where over 100 plants were recently counted. It perhaps also occurs on cliffs near Malin Head, East Donegal, but when the site was visited c.1970 a landslip had occurred and the species could not be found.

1/1 GB post 1970   1 SSSI   TN = 9 E
1/1   H post 1950   E in Europe

Armeria maritima (Mill.) Willd.   Thrift
subsp. elongata (Hoffm.) Bonnier

This inland subspecies of the common coastal species occurs on sandy grassland in Lincolnshire where seven colonies were reported in the 1950s. However, by 1970 four of these could no longer be found having been ploughed up or built upon. It will probably only survive in a Nature Conservation Trust reserve and within the boundary of a cemetery.

2/6 GB post 1970   1 NCTR : 3 SSSI   TN = 12 V

Primula scotica Hook.   Scottish Primrose

This endemic perennial herb is widespread in maritime turf in Sutherland, Caithness and Orkney: there is a single isolated locality in East Ross. Although threatened locally by oil developments in Orkney the total number of localities is such that the species is not in danger.

IUCN List

Cyclamen hederifolium Ait.   Cyclamen

As a possible native this perennial occurs now only in Kent though formerly it was also in Sussex. Only two or three stations in woodland are known at one of which plants were reported to have been dug up in 1967. It is also recorded occasionally as a garden escape in hedgebanks and woods elsewhere but there is no evidence that it becomes established in such places.

2/4 GB post 1950   2 SSSI   TN = 12 V

Centaurium tenuiflorum (Hoffmanns. & Link) Fritsch   Slender Centaury

This annual species of damp grassy places near the sea has been recorded only from two areas in Dorset and from the Isle of Wight. In the Dorset area hundreds of plants were recorded over three miles of coast in 1960: in the Isle of Wight 12 plants were seen in 1953 but despite frequent searches it has not been refound since. The species was once also in Guernsey and Jersey.

2/3 GB post 1960   2 NT   TN = 12 V
0/2   S post 1837

Centaurium latifolium (Sm.) Druce   Broad-leaved Centaury

This endemic annual of dune slacks was known in several places near Liverpool between 1804 and 1872. Its extinction may, in part, have been caused by collectors as many herbarium specimens are known.

0/1 GB post 1886   EXTINCT

35

Centaurium scilloides

Centaurium scilloides (L.f.) Samp.   Perennial Centaury

C. portense (Brot.) Butcher

This perennial herb occurs on grassy cliffs in Cornwall where two stations are known, in one of which it may not be native. In Pembrokeshire there are several adjacent stations where it is locally frequent and perhaps increasing. Generally, though no special protection is afforded, the species is not threatened except by mowing and by cutting at one station. It has, however, gone from one locality in Cornwall due to the habitat becoming completely overgrown by tall gorse during the past few years.

2/3 GB post 1960   1 SSSI   TN = 9 V

Gentiana verna L.   Spring Gentian

This perennial of stony grassy places on limestone in Yorkshire, Durham, Westmorland and Cumberland occurs exclusively at high altitudes between 1200 and 2500 ft. (380-800 m.). Also in western Ireland, in Clare, Galway and formerly Mayo, where it rises from sea-level to 1000 ft. (300 m.). It is apparently reasonably widespread and abundant, around 55 populations being known in Britain and at least 27 in Ireland, but it is declining in vigour in at least one station and is sparse and very local in certain outlying localities. At least 40 of the British populations are in National Nature Reserves, but the species still suffers greatly in places from deliberate and gross depredation by botanists and gardeners.

5/6 GB post 1960   40 NNR   TN = 6 R
20/26   H post 1930   Scheduled species

Gentiana nivalis L.   Alpine Gentian

This annual grows on rock ledges on mountains in Perthshire and Angus: there are about 10 colonies in some of which the species is very rare, for example one plant only in one, though in another, in 1964, the plants were too numerous to count. Whilst population fluctuations may occur naturally with an annual, this species has certainly declined in some places because of over-collecting and, elsewhere, is affected by heavy grazing. It is now one of the rarest British mountain plants and some sites are too easily accessible to collectors to be safe.

4/4 GB post 1960   5 NNR : 1 NT   TN = 7 V
                                Scheduled species

Gentianella anglica (Pugsl.) E.F.Warb.   Early Gentian

This endemic biennial herb of cliffs, dunes and grassland on chalk and limestone is now known from over 35 10 kilometre squares from Cornwall to Lincolnshire. Subsp. *cornubiensis* Pritchard is confined to two or three localities on cliffs on the north coast of Cornwall: the remaining localities are referable to subsp. *anglica* which, though apparently extinct in its single dune locality in North Devon, and showing some decline in central England, is generally so frequent as not to be in any danger.

IUCN List

Gentianella uliginosa (Willd.) Börner   Dune Gentian

This annual or biennial of dune slacks in Glamorgan, Carmarthenshire and Pembrokeshire still persists in small quantity in most of the eight localities from which it has been recorded, but in one locality at least it is threatened by increasing public pressure from horse-riders, and extensions to a golf course may have eliminated another.

5/5 GB post 1960   2 NNR : 2 SSSI   TN = 8 V
                                                          V in Europe

Polemonium caeruleum L.   Jacob's-ladder

This perennial herb is more or less confined to steep shaded slopes on Carboniferous limestone but is also occasionally found in woods. It is locally frequent in Staffordshire, Derbyshire and parts of Yorkshire and there is an outlying locality in Northumberland. A single record from Westmorland has not been recently confirmed. Other records are escapes from gardens of this commonly cultivated species. About 30 localities are known and there is no particular threat.

15/18 GB post 1950   1 NT : 5 SSSI   TN = 5 R

Cynoglossum germanicum Jacq.   Green Hound's-tongue

A biennial of woods and hedgebanks which was once widespread throughout central and southern England but is now confined to about ten localities in Kent, Surrey, Oxfordshire, Buckinghamshire and Gloucestershire. In one Oxfordshire locality over 1000 plants were recorded in 1967, but in another in Surrey there are only a few plants and it is still decreasing. The cause of the decline is not known. Perhaps of climatic origin but also in part due to changes in woodland management.

6/50 GB post 1950   1 NT : 2 SSSI   TN = 10 V

Myosotis alpestris

Myosotis alpestris Schmidt   Alpine Forget-me-not

This rhizomatous perennial occurs in upland limestone grassland in Westmorland and on mica-schist ledges and slopes in Perthshire: the English and Scottish populations are morphologically different. There are four stations in northern England and about seven in Scotland in some of which the species is very abundant: millions of plants have been assessed at one place. The English populations are subject to heavy grazing, the result of which has been the formation of a dwarf ecotype. In Scotland, the plants are mostly inaccessible to grazing sheep when on ledges but vulnerable on slopes below the cliffs.

6/7 GB post 1950   3 NNR : 2 SSSI   TN = 6 R

Buglossoides purpurocaerulea (L.) I.M.Johnston   Purple Gromwell
Lithospermum purpurocaeruleum L.

This creeping perennial is found in bushy places at the margins of woods on chalk and limestone in Devon, Somerset, Glamorgan and Denbighshire; formerly also in Kent and Monmouthshire as a native and recorded as a casual from eight scattered localities. Thirteen native stations are known in some of which thousands of plants occur but in others it is endangered because it is very attractive and the habitat is vulnerable. Protected to some extent by its early flowering.

11/22 GB post 1950   1 LNR : 5 SSSI   TN = 9 R

Echium plantagineum L.   Purple Viper's-bugloss
E. lycopsis L. pro parte

This erect biennial grows on cliffs and in arable fields on granite near the sea in Cornwall and the Isles of Scilly. It has long been known as a native in Jersey and as a casual on other Channel Isles. Occasional as a garden escape or a casual elsewhere. It appears to be very local and vulnerable to changes in farming practice.

4/6 GB post 1960   1 SSSI   TN = 9 V
1/1   S post 1970

Linaria supina (L.) Chazelles   Prostrate Toadflax

This prostrate annual is perhaps native in a few sandy places in Cornwall but is certainly naturalised elsewhere in that county on old railway lines, and in Devon and Carmarthenshire. Also recorded as a casual in Dorset, Isle of Wight, Hampshire, Glamorgan, Durham and the Channel Isles.

6/15 GB post 1950   No conservation   TN = 10 V

Scrophularia scorodonia L.   Balm-leaved Figwort

This perennial is locally abundant in hedgebanks, on roadsides, in marshes and among sand dunes in the Isles of Scilly, Cornwall and Devon. It is abundant in some of the Channel Isles and has recently been reported from an island off the coast of Pembrokeshire. Occasionally elsewhere as a casual. The cause of its apparent decline is unknown, but it may be under-recorded.

10/25 GB post 1950   1 SSSI   TN = 6 R
 3/3   S post 1970

Limosella australis R.Br.   Welsh Mudwort
L. subulata Ives

This annual occurs on wet mud at the edges of pools or where water has stood in Glamorgan, Merioneth and Caernarvonshire. There are eight populations in five main areas. Sometimes abundant, but it is apparently sensitive to small habitat changes which cause large fluctuations in the populations annually. Habitat protection is minimal and virtually impossible.

3/3 GB post 1960   1 NNR : 4 SSSI   TN = 5 R

Veronica spicata L. subsp. **spicata**   Spiked Speedwell

This creeping perennial of dry basic grassland is confined to the Breckland of East Anglia where four localities are now known in Suffolk, Norfolk and Cambridgeshire, some of which are large, containing several hundred shoots. Formerly more widespread but many colonies have been destroyed by ploughing or are threatened by tree planting.

3/10 GB post 1970   1 NNR   TN = 11 V
                          Scheduled species

Veronica fruticans Jacq.   Rock Speedwell

About 20 colonies of this perennial are known at present on mountains in Perthshire, Angus, Aberdeenshire and Inverness-shire. It was formerly reported from Banffshire and Argyll. It is very local and is seldom found in any abundance though there are several strong colonies. In one site it grows on ungrazed stony ledges where there is very little competition from other species. As it is an attractive species, though reasonably inaccessible, it is collected to some extent in at least one locality.

14/20 GB post 1950   8 NNR : 4 SSSI   TN = 5 R

Veronica verna

Veronica verna L. Spring Speedwell

An annual of open habitats in recently disturbed dry breck, which still occurs in some abundance in about half of the 19 stations known in the 1950s. It was last seen in its only station in Norfolk in 1946 where it has now been introduced into one Nature Conservation Trust reserve. Spraying, planting and lack of rabbit grazing and burrowing have been mainly responsible for its disappearance, but it is said to be in less danger of extinction than are *V. triphyllos* and *V. praecox* as it survives in less cultivated areas. It was once introduced in Devon.

1/8 GB post 1970 6 SSSI TN = 9 E

Veronica praecox All. Breckland Speedwell

An annual found only in Suffolk and Norfolk in a few cultivated fields where the soil is light and sandy. It may possibly be an introduction in all stations in this area as it was first recorded in Britain in 1933, and is certainly introduced in Oxfordshire. Spraying and other agricultural practices have been responsible for its decline. It has been introduced into one Nature Conservation Trust reserve.

3/3 GB post 1970 No conservation TN = 8 E

Veronica triphyllos L. Fingered Speedwell

This annual of sandy arable fields has been reported recently only in Surrey, Suffolk and Norfolk. Around 1950 at least ten localities were known, by 1970 this had been reduced to three and it now seems likely that only one locality remains and this is threatened by road-works. Changes in agricultural practice have been mainly responsible for this marked decline.

1/24 GB post 1970 No conservation TN = 12 E

Rhinanthus serotinus (Schönh.) Oborny Greater Yellow-rattle

This annual generally occurs in arable fields and waste places, and much less commonly on sandhills or in meadows. It was once widespread but is now known from only about ten localities in Surrey, Worcestershire, Lincolnshire, Yorkshire and Angus. However in one chalk grassland site in Surrey it occurs in great quantity and is in no danger of extinction there. Past misidentifications have perhaps given a false impression of its former range, whilst destruction of its vulnerable habitat is clearly responsible for its present limitation to so few localities.

5/68 GB post 1950 1 SSSI TN = 9 V

Melampyrum arvense L. Field Cow-wheat

An annual weed of cornfields now found in about six localities in Wiltshire,

Isle of Wight, Essex and Bedfordshire, in most of which it is very scarce. It was formerly much more widespread but has declined because of the drastic methods of modern agriculture: the use of sprays, removal of hedges and ditches, the burning of stubble and hedgerows. Populations fluctuate markedly: at one station there were 250 plants in the 1950s but only one in 1969, and none in 1973.

5/40 GB post 1970   No conservation   TN = 11 E

Euphrasia rhumica Pugsl.
This endemic annual is known only from two localities on the island of Rhum in the Inner Hebrides. The taxa is similar to *E. micrantha* Reichenb. and may be of hybrid origin.

1/1 GB post 1930   1 NNR   TN = 6 R

Euphrasia eurycarpa Pugsl.
This endemic annual is known from only three localities on the island of Rhum in the Inner Hebrides. The taxon is similar to *E. ostenfeldii* (Pugsl.) Yeo (*E. curta* auct., non (Fries) Wettst.) and may be of hybrid origin.

1/1 GB post 1930   1 NNR   TN = 5 R

Euphrasia campbelliae Pugsl.
This endemic annual is only known with certainty from about nine localities in heathy grassland near the sea in the Isle of Lewis, Outer Hebrides. It is possibly of recent hybrid origin: the morphology suggests the parents may have been *E. micrantha* Reichenb. and *E. marshallii* Pugsl.

7/7 GB post 1940   No conservation   TN = 6 R

Euphrasia rotundifolia Pugsl.
This endemic annual is confined to grassy headlands and sea cliffs in Sutherland, the Outer Hebrides, Orkney and Shetland. It is very closely related to *E. marshallii* Pugsl. with which it has a sympatric distribution.

9/14 GB post 1930   1 SSSI   TN = 7 R

Euphrasia marshallii Pugsl.
This endemic annual is known from over 26 localities in the Inner Hebrides, Ross, Sutherland, Caithness, Outer Hebrides, Orkney and Shetland, mainly among rocks and in cliff-top turf, but occasionally in salt-marshes and on wet peat. Although threatened locally by oil developments in Orkney and Shetland the total number of localities is such that the species is not in danger.   IUCN List

Euphrasia cambrica

Euphrasia cambrica Pugsl.

This endemic annual is known only from about six localities on grassy mountain slopes and cliffs in Caernarvonshire where it takes the place of *E. frigida* Pugsl. Records from Brecon and Westmorland have not been confirmed.

2/3 GB post 1950   1 NNR : 1 SSSI   TN = 7 R

Euphrasia heslop-harrisonii Pugsl.

This endemic annual is known only from marshes, maritime grassland and coastal cliffs in Inverness, the Inner Hebrides, Ross, Sutherland and Orkney where it is often locally abundant. It is possibly derived from *E. scottica* Wettst. by hybridization.

10/11 GB post 1950   1 NNR   TN = 6 R

Euphrasia rivularis Pugsl.

This endemic annual is characteristic of rocky stream banks and damp pastures in the mountains of North Wales and the Lake District.

9/11 GB post 1930   1 NNR : 2 SSSI   TN = 5 R

Euphrasia vigursii Davey

This endemic annual is confined to mixed heath in Cornwall and Devon where it is known from about 20 localities. There are many other records from the past and the apparent decline may be due to under-recording. Probably originated by hybridization between *E. anglica* Pugsl. and *E. micrantha* Reichenb.

13/25 GB post 1950   1 NNR : 1 NT   TN = 7 R

Bartsia alpina L.   Alpine Bartsia

About 20 colonies of this perennial, which occurs in upland meadows and on rock ledges on basic rocks at relatively high altitudes, are known from Yorkshire, Durham, Westmorland, Perthshire and Argyll. Only threatened in the southern part of its range where some of the populations are very small.

12/15 GB post 1950   2 NNR : 6 SSSI   TN = 4 R

Orobanche purpurea Jacq.   Yarrow Broomrape

A parasite of Yarrow, *Achillea millefolium* L., and certain other composites, now to be found only in Dorset, Isle of Wight, Kent, Norfolk

Pembrokeshire and Lincolnshire. It is still also in the Channel Isles. Formerly more widespread having been recorded from Devon (as an introduction), Somerset, Hampshire, Hertfordshire and Monmouthshire. Several of the extant populations are threatened by natural or human erosion, and one is on the edge of a municipal tip, where it is doomed to extinction in the near future.

8/19 GB post 1950   2 NT : 2 SSSI   TN = 8 V
3/4   S post 1970

Orobanche caryophyllacea Sm.   Bedstraw Broomrape

This parasite on Hedge Bedstraw, *Galium mollugo* L., has only been certainly recorded from Kent: there is doubt about the authenticity of a record from Argyll. Only two stations are now known neither with large populations: a roadside verge site was recently exterminated by road widening.

2/4 GB post 1960   2 SSSI   TN = 12 E

Orobanche reticulata Wallr.   Thistle Broomrape

This parasite on species of thistle, *Cirsium* and *Carduus*, is now known from only four localities in Yorkshire, where it was formerly more widespread. In one station a steady decline has been noticed since 1939, when there were over 400 spikes, until 1966 and 1967 when one and six spikes respectively were recorded. Rare in other stations where population sizes fluctuate from year to year. It is endangered by ploughing, spraying, dumping, gravel working and road widening. The Cheshire record has not been recently confirmed.

4/6 GB post 1950   No conservation   TN = 11 E

Orobanche loricata Reichenb.   Oxtongue Broomrape
O. picridis F.W. Schultz

This parasite on species of *Crepis* and *Picris* especially Hawkweed Oxtongue, *Picris hieracioides* L., is now known only from about six localities in Somerset, Kent and Buckinghamshire. It was formerly also recorded from Surrey, Hertfordshire, Oxfordshire, Suffolk, Cambridgeshire, Worcestershire, Brecon and Guernsey. Some of the extant localities are for single specimens and it is only locally abundant in one Kent station. The true decline is probably not as great as it appears, as some of the earlier records may have been misidentifications.

5/23 GB post 1950   1 SSSI   TN = 11 V
0/1   S post 1900

Orobanche maritima

Orobanche maritima Pugsl. Carrot Broomrape

This parasite which generally occurs on Wild Carrot, *Daucus carota* L., is now known only from Cornwall, Devon, Hampshire and Kent. It also occurs in three of the Channel Isles. Previously also recorded from Dorset, Suffolk, Glamorgan and Pembrokeshire. Populations are usually small in number, but plentiful in some years.

13/35 GB post 1950   1 NCTR : 1 LNR : 4 SSSI   TN = 8 V
4/5   S post 1970

Pinguicula alpina L. Alpine Butterwort

This bog species was first reported in its sole British locality in East Ross in 1831 and apparently became extinct just after 1900 because of the degeneration of the habitat following the growth of pines. Doubts have recently been expressed about the validity of the determination of the material in herbaria.

0/1 GB post 1909   **EXTINCT**

Mentha pulegium L. Pennyroyal

This creeping perennial of pond and lake margins and of damp grassland near the sea was formerly scattered throughout the lowlands of England and Wales where it was recorded from 55 of the the 71 vice-counties. It was also known from 13 out of 40 Irish vice-counties, mainly in the south but with a few localities around L. Neagh in the north, and from three of the Channel Isles. However it has declined rapidly in the first half of this century with improved drainage and the filling in of ponds or the cessation of goose and duck grazing round their margins. Since 1960 Pennyroyal has only been recorded from about 14 localities in Cornwall, Devon, Dorset, Hampshire, Sussex, Surrey, Berkshire, Brecon, Leicestershire and from Jersey in the Channel Isles. There are no recent records from Ireland.

14/∞ GB post 1960   1 FSCR   TN = 10 V
1/3   S post 1970
0/32   H post 1960

Thymus serpyllum L. Breckland Thyme

A perennial of very open, dry, sandy heaths, now found only in about 18 localities in the Breckland of Suffolk and Norfolk. Also recently refound on the chalk in Cambridgeshire. The sizes of all the extant populations of this critical taxon are not known, but some are extensive.

5/6 GB post 1960   1 NNR : 3 SSSI   TN = 6 R

44

Calamintha sylvatica Bromf.   Wood Calamint

The only known locality for this perennial in Britain is a chalky bank on the Isle of Wight where, since 1900, and for no obvious reason, the population had become very small. A nature reserve has now been established for its protection and it is not therefore endangered.

1/1 GB post 1970   1 NCTR   TN = 8 V

Salvia pratensis L.   Meadow Clary

This perennial is probably native in calcareous grassland in about a dozen localities in Wiltshire, Kent, Surrey, Berkshire, Oxfordshire, Gloucestershire and Monmouthshire: formerly also in Rutland. It is not infrequent as an established introduction and as a casual in waste places. Most of the native populations are small and on private land, where they are endangered by ploughing, over-grazing or fertilizing.

14/28 GB post 1950   1 NCTR   TN = 10 V

Stachys germanica L.   Downy Woundwort

This attractive perennial species of calcareous pastures and roadside verges is now believed to be extinct except in five localities in Oxfordshire. Most of these populations have up to 20 plants but numbers vary widely from year to year. Formerly also recorded from Hampshire, Northamptonshire, Denbighshire and Lincolnshire, and as an introduction from several localities in the same general area. There is a doubt about the authenticity of the Denbighshire record (unsupported by a specimen) following the recent discovery of *S. alpina* L. in a wood adjacent to the site from which *S. germanica* was originally reported.

2/10 GB post 1960   No conservation   TN = 13 E

Stachys alpina L.   Limestone Woundwort

This attractive perennial was known until recently from only a single locality in Gloucestershire where it was threatened by overgrowth by rank grasses and the planting of conifers. The area has, however, now become a Nature Conservation Trust Reserve and the future of the plant is more secure. At one time eight of the colonies were known in Gloucestershire and another in Denbighshire. This last site has been destroyed by road widening, but in 1973 a second site was found in the area and two healthy clumps are reported.

2/3 GB post 1970   1 NCTR   TN = 12 E

Galeopsis segetum

Galeopsis segetum Neck.   Downy Hemp-nettle

An annual weed of arable land which was recorded as a native only in Caernarvonshire, and as a casual, before 1930, in Kent, Essex, Suffolk, Carmarthenshire, Lincolnshire, Nottinghamshire, Derbyshire, Cheshire, Yorkshire and Northumberland, and more recently in Devon. Appearances in its native station were always erratic, being controlled by the amount of ploughing. It was recorded in 1957, but following the cessation of all arable cultivation was not seen again until 1975.

1/24 GB post 1957   No conservation   TN = 12 E

Teucrium scordium L.   Water Germander

This perennial herb occurs on the banks of rivers, pits and ditches on calcareous soils, and in dune slacks. Since 1950 it has only been recorded from Devon, Cambridgeshire and Lincolnshire but it is now almost certainly extinct in the last named county. It is now known from two stations and from a third where it has recently been reintroduced after being lost for several years. Formerly also recorded from Berkshire, Oxfordshire, Suffolk, Norfolk, Huntingdonshire, Northamptonshire and Yorkshire. It is still abundant in some parts of western Ireland but appears to be extinct in the Channel Isles. Most of the losses have been due to drainage and reclamation of fenland sites.

3/22 GB post 1970   1 NNR : 2 NCTR   TN = 10 V
0/2   S post 1926
8/19   H post 1930

Teucrium botrys L.   Cut-leaved Germander

About eight localities for this annual or biennial are known from fallow fields, open grassland and woodland clearings on the chalk in Hampshire, Kent, Surrey and Gloucestershire. Previously known also from Wiltshire but it became extinct there after alteration of the habitat. At one Surrey station it was present in quantity when the field was arable but decreased as a sward developed and is now only maintained by rotovating one bare patch where, in 1970, 43 plants survived. One station in the county was lost because of building, whilst another became a nudist colony where no reputable botanist has seen it recently.

8/11 GB post 1950   1 NNR : 1 NCTR : 1 SSSI   TN = 8 R

Ajuga genevensis L.

This introduced perennial which was formerly established in two localities in Cornwall and Berkshire seems to have become extinct in both during the last ten years.

0/2 GB post 1967 **EXTINCT**

Campanula persicifolia L. Peach-leaved Bellflower

This perennial herb may have been indigenous in Devon, Surrey, Berkshire and Gloucestershire, but it is now believed to be extinct as a native in all of these. It has occurred occasionally as an introduction or garden escape but rarely persists.

0/5 GB post 1949 **EXTINCT**

Campanula patula L. Spreading Bellflower

This biennial to perennial herb of shady banks and woodland margins was formerly recorded as a native from 22 vice-counties in England and Wales from Hampshire and Kent to Shropshire and South Yorkshire. It was recorded from 26 10 kilometre squares during the period 1930-1960 which was the basis of the map in the *Atlas of the British Flora*. Since 1960 the only records are from Surrey, Berkshire, Gloucestershire, Herefordshire, Warwickshire, Shropshire and Radnorshire, though it seems likely that it still persists in Monmouthshire. The cause of this decline is not known.

14/91 GB post 1960 No conservation TN = 11 V

Campanula rapunculus L. Rampion Bellflower

This introduced biennial was formerly widely naturalised in fields and hedgebanks in England and South Scotland where it was a relic of its cultivation as a winter salad vegetable. Now this vegetable is less commonly grown, records are few and it appears to be established only in Hampshire, Surrey, Essex and Berkshire.

7/75 GB post 1960 No conservation TN = 12 V

Phyteuma spicatum L. Spiked Rampion

This handsome perennial herb of woods, wooded roadside verges and steep banks is confined to Sussex where it is known from at least ten localities. It flourishes in some woods but suffers from mowing in some roadside verge stations, despite their scheduling as verge reserves. However the species is in no sense endangered. Also recorded as a garden escape from Warwickshire, Staffordshire, Merionethshire, Derbyshire and Roxburghshire.

6/6 GB post 1960 6 NCTR TN = 8 R

Lobelia urens

Lobelia urens L.   Heath Lobelia

This attractive perennial of rough pastures or grassy heaths, and at the margins of woods on damp and infertile acid soils, is known from about ten localities in Cornwall, Devon, Dorset, Hampshire and Sussex. It was formerly also in Kent and Herefordshire. It is rather rare in Cornwall, but more flourishing populations are reported from Devon and Hampshire. It is very vulnerable to grazing by cattle but disappears when the community becomes closed, particularly if bracken is present. It benefits when the habitat is somewhat disturbed, especially in damp places.

8/12 GB post 1960   1 SSSI   TN = 9 V

Galium fleurotii Jord.

This taxon has only recently been recognised as occurring in Britain. It is a segregate of *G. pumilum* Murr., and is confined to the Cheddar area of North Somerset where a large population occurs.

1/1 GB post 1960   1 SSSI   TN = 9 R
                                                                    R in Europe

Galium debile Desv.   Slender Marsh-bedstraw

This perennial herb of damp acid hollows, pond margins and ditch banks still occurs in about a dozen localities in Devon and the New Forest of Hampshire. It has recently been reported from a single isolated locality in Yorkshire but the record has not been substantiated. Formerly also in the Channel Isles but apparently now extinct there.

6/6 GB post 1950   3 SSSI   TN = 6 R
0/2   S post 1924

Galium tricornutum Dandy   Corn Cleavers

This annual cornfield weed was once widespread throughout lowland England as far north as Northumberland and was occasionally recorded in Wales and Scotland: it was probably native in 54 vice-counties. Although some decline took place before 1930 this species was still locally abundant, especially in fields on the chalk and the oolite, until the 1950s. However since 1960 it has been recorded only from Wiltshire, Hampshire, Sussex, Essex, Berkshire, Cambridgeshire, Bedfordshire and Yorkshire and only two of these records (Cambridgeshire and Bedfordshire) date from the 1970s. The pattern of decline closely resembles that of *Agrostemma githago* and is doubtless related to the same combination of causes — cleaner seed corn and the use of herbicides.

13/ ∞ GB post 1960   No conservation   TN = 10 E

Galium spurium L.   False Cleavers

This doubtfully native annual herb of arable fields, allotments and waste ground may now be extinct in all its long-established localities in Somerset, Cambridgeshire and Staffordshire. However it was recorded as recently as 1972 in Essex in the area where it had been first recorded nearly 130 years earlier. A sporadic species of this kind could well reappear elsewhere. Also occasionally reported as a casual.

1/4 GB post 1960 No conservation TN = 13 E

Lonicera xylosteum L. Fly Honeysuckle

This small deciduous shrub is confined as a native to a single locality on the chalk downs of Sussex where one extensive colony occurs. Here, though rare, it is still apparently reasonably safe. It is widespread elsewhere as an introduction occurring in many places in England and Wales and a few in Scotland and Ireland.

1/1 GB post 1960 No conservation TN = 11 V

Valerianella rimosa Bast. Broad-fruited Cornsalad

An annual cornfield weed formerly widespread in England, Wales and Ireland, mainly in the south and east, and only recorded from Fife in Scotland: it was reported from 47 vice-counties in Great Britain and 21 in Ireland. However it appears to have declined rapidly: whilst there were some losses in the nineteenth century the major changes have been more recent. In the period 1930-1960 it was reported from 43 10 kilometre squares whereas, since 1960, it has been found only in nine squares in Dorset, Hampshire, Sussex, Berkshire, Bedfordshire and Leicestershire. The reason for this decline is unknown but may be related to changes in agricultural practice. However this species is easily overlooked and other extant localities may have remained unreported.

9/96 GB post 1960 No conservation TN = 9 V
9/41 H post 1930

Valerianella eriocarpa Desv. Hairy-fruited Cornsalad

This introduced annual weed of banks, dunes, old walls, arable fields and other dry open habitats, has been reported recently from a few localities in Cornwall, Dorset, Middlesex, Northamptonshire, Gloucestershire, Leicestershire and Moray, and from the Channel Isles. It was formerly also reported from Devon, Somerset, Wiltshire, Isle of Wight, Hampshire, Kent, Surrey, Hertfordshire, Bedfordshire, Shropshire, Pembrokeshire, Caernarvonshire, Derbyshire, Yorkshire and the Lothians, but in most of these it was casual only and did not persist.

10/31 GB post 1950 No conservation TN = 8 V
1/2 S post 1950

Senecio cambrensis

Senecio cambrensis Rosser

This endemic is a winter annual which may have arisen as the result of hybridization between *Senecio squalidus* L. and *S. vulgaris* L. and was described by Rosser who first discovered it in Flintshire in 1953, where it still persists. Three localities are known from Denbighshire and recent records from Shropshire suggest that this species may now be spreading. Observation suggests that it does not tolerate competition but that in open habitats such as roadsides and gardens it sets seed and persists for many years.

5/5 GB post 1970   No conservation   TN = 9 R

Senecio paludosus L.   Fen Ragwort

This species was formerly recorded from numerous fen ditches in Suffolk, Norfolk, Cambridgeshire and Lincolnshire. It apparently became extinct towards the end of the nineteenth century as a result of the drainage of the fens, but a small population was found in Cambridgeshire in 1972.

1/8 GB post 1970   No conservation   TN = 13 V

Senecio congestus (R.Br.) DC.   Marsh Fleawort
S. palustris (L.) Hook.

This biennial or sometimes perennial herb was once widespread in fen ditches in the eastern counties of Suffolk, Norfolk, Cambridgeshire, Huntingdonshire, Lincolnshire and Yorkshire, with an isolated locality in Sussex. However its habitats were destroyed by the drainage of the fens and it became extinct before the end of the nineteenth century.

0/24 GB post 1899   **EXTINCT**
V in Europe

Homogyne alpina (L.) Cass.   Purple Colt's-foot

This perennial herb is recorded only from Angus and the Outer Hebrides. In Angus a small population of at least 20 plants covers a few square metres at an altitude of over 2000 ft. (600 m.): the size of the Hebridean population is not known. It was first found in Angus around 1800 but was not refound until 1951. The true status of this species is uncertain and it is believed by some to be an introduction. Saved to some extent by the similarity of its leaves to *Tussilago farfara* L.

2/2 GB post 1950   1 NNR   TN = 7 R

Pulicaria vulgaris Gaertn.   Small Fleabane

This annual herb grows in moist sandy places or on pond margins and in similar places where water stands in winter. It has been decreasing rapidly being found now only in Wiltshire, Hampshire, Sussex and Surrey whereas, previously, it had been recorded from about 30 other, mainly southern, vice- counties as well as the Channel Isles. As far as is known there are now only nine extant populations all of which are small: only six plants were recorded in one Surrey station in 1962 and these have since disappeared. However several new localities have been found recently in the New Forest where it has apparently spread on ponies' hooves. Drainage, natural succession, and a falling off in the keeping of geese and ducks which kept down the grass and enriched the habitat are mainly responsible for its general decline and many of the remaining stations are endangered.

9/117 GB post 1950   No conservation   TN = 11 V
0/3   S post 1877

Filago lutescens Jord.   Red-tipped Cudweed

F. apiculata G.E.Sm.

This annual weed of sandy arable fields, gravel pits and banks was formerly widespread in southern and eastern England: it was known from 22 vice-counties extending west and north to Worcestershire and South Yorkshire. However it has declined rapidly and is now known only from Surrey, Essex, Suffolk and Norfolk where, since 1960, it has been reported from about 15 localities though it is already believed to have become extinct in some of these. The cause of the decline is unknown but is doubtless connected with changes in arable farming practice.

10/67 GB post 1960   No conservation   TN = 9 V

Filago pyramidata L.   Broad-leaved Cudweed

F. spathulata Presl

This arable weed of sandy fields and open disturbed habitats formerly occurred in 25 vice-counties in southern and eastern England from Cornwall to Lincolnshire, and in Jersey in the Channel Isles. However it has declined rapidly and since 1960 has been reported from only 11 localities in Hampshire, Kent, Surrey, Hertfordshire, Berkshire, Norfolk and Cambridgeshire. The cause of this decline is uncertain, but is presumably related to changes in agricultural practice.

11/96 GB post 1960   No conservation   TN = 9 V
0/1   S post 1931

51

Filago gallica

Filago gallica L.   Narrow-leaved Cudweed

This introduced annual was first recorded in Britain in 1696 and became established in several dry, grassy places in Hampshire, Kent, Surrey, Essex, Hertfordshire, Middlesex, Suffolk and the Channel Isles. It appears to have become extinct in Britain about 1955 though, its occurrence being sporadic, it may reappear. In one Essex locality, where it persisted for over 100 years, the gravel pit in which it occurred gradually became overgrown through lack of disturbance. Apparently it still occurs in Sark in the Channel Isles.

0/10 GB post 1955   **EXTINCT**
1/2   S post 1970

Gnaphalium norvegicum Gunn.   Highland Cudweed

This rare perennial herb only occurs at high altitudes up to 3600 ft. (1100 m.) in Scotland, where it is currently known from eight localities in Angus, Inverness-shire and Ross-shire. It was formerly also recorded in Perthshire, Aberdeenshire and Moray. While found only in small quantity on a few mountains with only solitary rosettes, a strong colony is reported from Inverness-shire. Although overall it is local and rare all populations are thought to be stable except one threatened by the development of a ski-slope.

9/14 GB post 1950   3 NNR : 4 SSSI   TN = 6 R

Gnaphalium luteoalbum L.   Jersey Cudweed

An annual of sand dunes, sandy fields and waste places, possibly an introduction in most British stations, but almost certainly native in Norfolk and the Channel Islands. It still occurs in the Channel Islands but in Norfolk where 40-50 plants were recorded in 1967 it has not been seen since 1973. Formerly also in Suffolk and Cambridgeshire but now almost certainly extinct there. The decline has almost certainly been due to improvements in agriculture. Also recorded occasionally as a casual.

1/5 GB post 1970   No conservation   TN = 12 E
2/2   S post 1950

Aster linosyris (L.) Bernh.   Goldilocks Aster
Crinitaria linosyris (L.) Less.

This woody perennial herb is confined to coastal limestone cliffs in Devon, Somerset, Glamorgan, Caernarvonshire and Lancashire: there are now about seven colonies, one other having become extinct. Also recorded as a casual in Hampshire, Sussex, Leicestershire, Yorkshire, West Lothian and the Channel Isles. It is safe where inaccessible on cliffs as, for example, in Lancashire, but is vulnerable elsewhere and has suffered from collecting. Some of the populations are very small and no extensive colonies are known.

6/7 GB post 1950   1 NNR : 1 LNR : 4 SSSI   TN = 5 R

Erigeron borealis (Vierh.) Simmons   Alpine Fleabane

This perennial herb of mountain rock ledges is now found only in Perthshire, Angus and Aberdeen, though formerly also reported from Banffshire and Inverness-shire. In most of its ten or so colonies the populations are limited to a few plants, though it is frequent in two stations. No immediate threat is known: it is protected by the inaccessibility of the habitat and by almost all the localities being in National Nature Reserves.

6/9 GB post 1950   13 NNR : 1 SSSI   TN = 5 R

Otanthus maritimus (L.) Hoffmanns. & Link   Cottonweed

This perennial maritime herb of fixed dunes and fine shingle was first recorded in Great Britain, in Essex, in 1597 and subsequently in the Isles of Scilly, Cornwall, Devon, Dorset, Hampshire, Kent, Suffolk, Caernarvonshire, Anglesey, the Channel Isles and Ireland. Now, however, it is extinct in Britain and the Channel Isles and is confined to one locality in Wexford in Ireland. It is a Mediterranean species at the northern edge of its range and it seems likely that the main cause of extinction has been a change in climate.

0/18 GB post 1968   **EXTINCT**
0/3   S post 1926
1/5   H post 1970

Artemisia norvegica Fr.   Norwegian Mugwort

This high montane species is known from only three localities at about 2500 ft. (800 m.) in Ross in the north of Scotland. In West Ross, where it was found for the first time in Britain in 1951, it is more abundant than in the other two sites. When first found about 24 colonies were counted and no decline has since been observed. All the localities have some measure of protection and collecting is the only obvious threat.

3/3 GB post 1960   1 NNR : 1 SSSI   TN = 5 R

Artemisia campestris L. Field Wormwood

This perennial herb is now confined to one extensive roadside and waste land site in Suffolk and to one area in the Breckland of Norfolk: it has long been extinct in Cambridgeshire. Recorded also as a casual in Cornwall, Hampshire, Glamorgan, Yorkshire, Durham, Renfrewshire and Fife. The Suffolk population has been reduced drastically in recent years by building: a population in Norfolk was eliminated by chicken farming, and the existing population is vulnerable to rabbit grazing. Transplants have been made to several Breckland nature reserves and over-zealous transplanting by well-wishers may now constitute its greatest threat.

2/11 GB post 1970   No conservation   TN = 11 E

Cirsium tuberosum (L.) All. Tuberous Thistle

A perennial herb of chalk downs and other calcareous pastures recently recorded from less than 20 localities in Wiltshire, Cambridgeshire and Glamorgan. Whilst abundant at one station in Wiltshire and on the Glamorgan coast, it is scarce elsewhere, only three plants having been recorded in Cambridgeshire in 1960 and ten in 1970 and there are fears that this population may now be extinct. The major cause of the continuing decrease and disappearance of the species is ploughing and, as in Wiltshire, hybridization with *Cirsium acaule* Scop.

11/13 GB post 1970   1 NNR : 1 NCTR   TN = 6 R

Centaurea calcitrapa L. Red Star-thistle

This biennial herb has been widespread as a casual in south and east England and occasionally elsewhere. It is now less frequent than formerly and has been recorded from only 14 10 kilometre squares since 1960. However it still persists in several localities on sandy or chalky soils along the Sussex coast where it was first recorded in 1765 and where it may possibly be a native species.

14/109 GB post 1960   1 SSSI   TN = 8 R
  0/2   S post 1925

Arnoseris minima (L.) Schweigg. & Koerte Lamb's Succory

This small annual of arable fields on sandy soil has only been seen since 1953 in about a dozen localities in Surrey, Berkshire, Buckinghamshire, Norfolk, Lincolnshire, Nottinghamshire and Yorkshire, but is has not been refound in Norfolk, Lincolnshire or Nottinghamshire since 1960 and is now almost certainly extinct in these three counties. There are no records from anywhere since 1970 and it seems likely that this species is almost

extinct in Britain. At two stations in Surrey the populations have decreased from many plants some years ago when the fields were arable to
none in 1970 when pasture had replaced the arable: this has probably also
been a cause of decline elsewhere.

4/76 GB post 1960   No conservation   TN = 10 E

Hypochoeris maculata L.   Spotted Cat's-ear

A perennial herb of calcareous pasture and sea cliffs occurring in about 13
stations in Cornwall, Hertfordshire, Suffolk, Cambridgeshire, Bedfordshire, Northamptonshire, Caernarvonshire, Lincolnshire and Lancashire,
and also on Jersey in the Channel Isles. However the Lincolnshire site has
not been refound recently, and it became extinct in Essex about 1850.
Nevertheless it has increased over the past ten years in a Bedfordshire
station, where it has been known since 1913 and where flowering seldom
occurred until regular sheep grazing ceased in 1930.

9/16 GB post 1970   2 NNR : 5 SSSI   TN = 6 R
1/1   S post 1970

Scorzonera humilis L.   Viper's-grass

This very rare perennial herb is known only from marshy fields in Dorset
where at one of the two stations it has recently become extinct. In Warwickshire, where it was possibly introduced with seed, it was locally
frequent when first recorded in 1954 but gradually became scarce and has
not been seen since 1965. The only remaining population in Dorset,
consisting of many thousands of plants, is now a nature reserve.

1/3 GB post 1970   1 NCTR   TN = 10 V

Lactuca saligna L.   Least Lettuce

This annual or biennial herb once grew in salt-marshes, on shingle, on sea
walls and in waste places, usually near the sea, in Sussex, Kent, Essex,
Middlesex, Suffolk, Norfolk, Cambridgeshire and Huntingdonshire.
Although it became extinct in Middlesex in 1800 it persisted in one or two
localities in the other counties until the 1950s. However since 1960 it has
only been observed at eight sites, in Sussex, Kent and Essex and a survey of
these sites in 1975 suggests that it may now survive only at a single Kent
locality where over 300 plants were observed. The cause of this decline is
not clear, though in some instances it is connected with work to improve
our sea defences.

4/32 GB post 1960   No conservation   TN = 11 E

Cicerbita alpina (L.) Wallr.   Alpine Sow-thistle

About ten colonies of this tall perennial herb are known, most being in rather inaccessible places on moist rocks on mountains in Angus and Aberdeenshire, none being really large: it has become extinct at one other station. In one Aberdeenshire colony where there were originally about 200 plants in the early 1960s there are now very few, and the colony has also lost vitality. Though collecting is not thought to be the cause here it has damaged other populations. Damage is also caused by deer grazing.

4/4 GB post 1950   2 NNR : 2 SSSI   TN = 5 R
                         Scheduled species

Crepis foetida L.   Stinking Hawk's-beard

This annual or biennial herb occurs on waysides and waste places, especially on chalk and shingle. Only three or four colonies have been reported recently from Sussex and Kent, though it may now be extinct in the former. It appears to have been more widely distributed in the past, having been recorded from Somerset, Essex, Middlesex, Berkshire, Suffolk and Bedfordshire, though it is possible that some of these records were errors. Nevertheless decline has undoubtedly occurred though the cause is unknown. Whilst over 400 flowering plants were recorded in one Kent colony in 1969, only two were seen in 1970.

2/19 GB post 1960   1 SSSI   TN = 9 V

Taraxacum glaucinum Dahlst.

Confined to sand-dunes on Anglesey and in north Lancashire, apparently in very small numbers. It has been lost from two sites in south Lancashire due to building.

2/4 GB post 1970   1 NNR   TN = 9 V

Taraxacum acutum A.J. Richards

An endemic species of short calcareous turf known from one site in Hertfordshire which has about 80 plants where it is threatened by golfing and collecting. Little is known about its other site in Norfolk.

2/2 GB post 1960   No conservation   TN = 9 V

Taraxacum austrinum Hagl.

This occurs in very small quantity at one site in Cambridgeshire and is also to be found in a fen in Guernsey, where there are at least 150 plants. It was formerly known from another site in Cambridgeshire and one in Jersey, but is now apparently extinct in both. It may still persist in Galway, in Ireland, but the exact locality is unknown.

1/2 GB post 1970   1 NCTR   TN = 9 V
1/2 S post 1970
1/1 H post 1950

Taraxacum cymbifolium Lb.f.

A high-arctic species, with its next nearest site in north Iceland, which occurs on a single Perthshire mountain, where there are at least 150 plants. This species may be in danger of over collection.

1/1 GB post 1970   1 NCTR   TN = 6 R

Taraxacum ziphoideum Hagl.

Three colonies of this species are known: two are in Inverness-shire where one of them is reported to be thriving but the other small; the third colony is in Argyllshire where only a single plant has been found. Outside Britain it occurs locally in Norway.

3/3 GB post 1970   No conservation   TN = 7 R

Taraxacum pseudonordstedtii A.J. Richards

This endemic species is locally common in a limited area of calcareous flushes in Upper Teesdale, Durham.

1/1 GB post 1970   1 NNR   TN = 5 R

Taraxacum hygrophilum van Soest

This species is locally frequent in a marshy basin in Kent. Otherwise only known from N. Holland, where it is very local.

1/1 GB post 1970   1 NCTR   TN = 6 R

Alisma gramineum Lejeune   Ribbon-leaved Water-plantain

An aquatic perennial now known from four localities in Norfolk, Cambridgeshire, Worcestershire and Lincolnshire. It was first recorded in Worcestershire in 1920. The wide scatter of the records and their recent origin suggest an introduction into Britain perhaps by migrating wildfowl from Denmark and other countries bordering the Baltic, where the species occurs as a native. This possibility is supported by the determination of some Norfolk material as subsp. *wahlenbergii* Holmb. known only from round the Gulf of Bothnia and the northern side of the Gulf of Finland. Threatened only by dredging of the rivers and dykes in which it grows.

4/4 GB post 1970   1 NNR : 1 SSSI   TN = 8 R

Damasonium alisma

Damasonium alisma Mill.   Starfruit

This annual herb is rapidly approaching extinction. Once recorded from about 50 10 kilometre squares in Hampshire, Sussex, Kent, Surrey, Essex, Hertfordshire, Middlesex, Berkshire, Buckinghamshire, Worcestershire, Shropshire, Leicestershire and Yorkshire, it was reduced to only six squares by 1950. Since 1970 it has only been reliably reported from a single locality in Surrey. It grows on the muddy margins of acid ponds, a habitat which has been destroyed by in-filling.

1/50 GB post 1970   No conservation   TN = 13 E

Sagittaria rigida Pursh   Canadian Arrowhead

This aquatic perennial is naturalised in the Exeter Canal in Devon where it has become well established over a short stretch since it was first recorded in 1898. Formerly also in the River Exe, but not seen there for about 30 years.

2/2 GB post 1970   2 SSSI   TN = 10 V

Hydrilla verticillata (L.f.) Royle

This submerged aquatic monocotyledon was first discovered in Esthwaite Water in Westmorland in 1915, where it has not been seen since 1934 and is now, apparently, extinct. It still occurs in a single station in Galway, western Ireland.

0/1 GB post 1934   **EXTINCT**
1/1   H post 1970

Scheuchzeria palustris L.   Rannoch-rush

This perennial herb of very wet *Sphagnum* bogs is now apparently confined to two squares in Perthshire, where in one of them however 400-500 flowering spikes have recently been counted. It was formerly also in Shropshire, Cheshire, Yorkshire, Inverness-shire and Argyllshire, but has been eliminated through a combination of drainage, peat cutting and afforestation. First located in Ireland in 1951 but now extinct there also, following drainage of the bog in which it occurred.

2/9 GB post 1970   1 NNR   TN = 12 E
0/1   H post 1960

Potamogeton nodosus Poir.   Loddon Pondweed

This aquatic herb of slow-flowing base-rich rivers has been recorded only from the Avon (Somerset, Wiltshire and Gloucestershire), Stour (Dorset),

Thames (Berkshire, Oxfordshire and Buckinghamshire), and Loddon (Berkshire). However there are no post-1950 records from the River Thames and it may have become extinct in this river.

7/11 GB post 1950   No conservation   TN = 7 R

Potamogeton epihydrus Raf.   American Pondweed

This pondweed is native only in three lochs in the Outer Hebrides where, though its presence has not been recently confirmed, it probably still persists. It is also known in small quantity as an introduction in five localities in canals in Lancashire and Yorkshire.

2/2 GB post 1945   1 SSSI   TN = 5 R
      R in Europe

Potamogeton rutilus Wolfg.   Shetland Pondweed

Until 1960 this submerged pondweed was known only from lochs in the Outer Hebrides and Shetland. However it has recently been reported from Inverness-shire and Ross on the mainland of Scotland as well as from the Inner Hebrides. Though the sizes of its populations are not generally known they are said to be locally abundant in one area. Probably spreading or overlooked.

9/9 GB post 1960   No conservation   TN = 4 R

Najas flexilis (Willd.) Rostk. & Schmidt   Slender Naiad

This submerged aquatic annual herb occurs in lakes in Kirkcudbrightshire, Perthshire, Kintyre and in the Inner and Outer Hebrides. Previously known from north Lancashire, but not recorded there for 40 years. It is also widespread but local near the west coast of Ireland. There is no evidence that any of the extant sites are endangered.

12/15 GB post 1950   2 SSSI   TN = 2 R
8/10   H post 1930

Najas marina L.   Holly-leaved Naiad

This submerged species of slightly brackish water is confined to the Norfolk Broads where, recent evidence suggests, it is declining in Broads open to boat traffic: in several localities where it was previously abundant it can no longer be found. Increased turbulence of the shallow water in which it grows caused by power boats may be a contributory factor.

2/3 GB post 1970   1 NNR : 1 NCTR : 1 SSSI   TN = 8 V

Eriocaulon aquaticum

Eriocaulon aquaticum (Hill) Druce   Pipewort
E. septangulare With.

This perennial herb occurs in shallow water in acid lochs where it often forms dense mats. It has been recorded in Inverness-shire and the Inner Hebrides from Coll to Skye. It is locally abundant in some parts of Skye where many colonies have been reported. It is also abundant in many places in western Ireland.

7/9 GB post 1960   1 SSSI   TN = 4 R
37/52 H post 1930

Polygonatum verticillatum (L.) All.   Whorled Solomon's-seal

This perennial herb of mountain woods in Perthshire may now be confined to only four or five localities. It was formerly also in Northumberland and Angus but has long been extinct in both these counties. The reason for these losses is unknown but may be due to natural alteration of the habitat.

4/10 GB post 1960   2 SSSI   TN = 9 V

Maianthemum bifolium (L.) Schmidt   May Lily

This slender creeping perennial occurs in acid woods in Norfolk, Lincolnshire, Yorkshire and Durham. Although local, in several localities over five thousand plants have been recorded. In some areas, e.g. Norfolk, it is regarded as an introduction, especially where associated with exotic conifers, but in others, especially in the north of England, it appears to be associated with entirely native woodland species. Formerly also recorded from Middlesex, Oxfordshire, Bedfordshire, Lancashire, Northumberland and Midlothian; its decline has been due, mainly, to the destruction of woodland by felling.

5/12 GB post 1960   1 SSSI   TN = 8 V

Asparagus officinalis L.   Wild Asparagus
     subsp. **prostratus** (Dumort.) E.F.Warb.

This creeping perennial has only been recorded recently from grassy sea-cliffs in Cornwall and Pembrokeshire. It also still occurs on two of the Channel Isles and in several coastal areas in south-east Ireland. Populations of 20 to 30 plants are not infrequent. Apparently declining, and no longer to be found in Somerset, Dorset, Gloucestershire, Glamorgan or Anglesey where it was known in the nineteenth century. Subsp. *officinalis* is widespread as an escape from cultivation.

4/12 GB post 1970	1 NNR : 1 NT : 2 SSSI	TN = 10 V
2/2 S post 1970		
5/6 H post 1930		

Fritillaria meleagris L.   Fritillary

A bulbous perennial of old alluvial meadows in southern England and the Midlands. Before 1930 it was present in 116 10 kilometre squares in 27 counties. By 1970 it was found, only in any quantity, in 15 squares in nine counties: Wiltshire, Berkshire, Oxfordshire, Buckinghamshire, Suffolk, Huntingdonshire, Gloucestershire, Herefordshire and Staffordshire. Though at a few sites thousands of plants occur most have less than one hundred. Whilst picking and grazing during the flowering season prevent regeneration by seed and reduce populations locally, the main cause of the dramatic decline of this species has been the drainage, ploughing and artificial fertilisation of the riverside meadows in which it occurred.

15/116 GB post 1970   1 NNR : 2 NCTR : 12 SSSI   TN = 9 V

Lloydia serotina (L.) Reichb.   Snowdon Lily

This bulbous perennial occurs on basic rock ledges on mountains only in Caernarvonshire where five populations are still extant in the more inaccessible parts. Collecting has apparently caused its disappearance elsewhere: it was formerly reported from 12 cliffs. Its occurrences are as solitary plants or, more rarely, as small groups some of which consist of 100 or more plants. The species has long since disappeared from the peak of Snowdon, and has noticeably declined in amount in another locality during the past 20 years.

2/2 GB post 1970   2 NNR   TN = 8 V
                              Scheduled species

Gagea saxatilis Koch

This bulbous perennial was not recognised as occurring in Britain until 1974, though it was first found in 1965 when, in the absence of adequate flowering material, it was identified as *Lloydia serotina*. The grass-like leaves and rare flowering of this species at the edge of its range has probably caused it to be overlooked for so long, because the habitat and associated species suggest that it is a native rather than a recent introduction.

1/1 GB post 1975   1 SSSI   TN = 10 E

Muscari atlanticum

Muscari atlanticum Boiss. & Reut. Grape Hyacinth

This bulbous perennial occurs in dry grassland and in hedgebanks in Suffolk and Cambridgeshire, mainly in the Breckland: it appears to be extinct in Norfolk. Records from elsewhere were usually errors for other naturalised species. Roadworks and ploughing have contributed to its noticeable decline though there is some evidence that at many sites it does not appear every year.

10/17 GB post 1950 No conservation TN = 9 V

Juncus dudleyi Wiegand

This introduced perennial rush is well established in marshy ground in Perthshire and Rhum in the Inner Hebrides, though the widening of a roadside verge may have endangered the Perthshire station.

2/2 GB post 1930 1 NNR TN = 5 R

Juncus filiformis L. Thread Rush

This wiry perennial grows on the stony margins of lakes and reservoirs in Leicestershire, Yorkshire, Durham, Westmorland, Cumberland, Fife, Stirlingshire and Kincardineshire. Though it has not been refound recently in several localities the species in general is spreading: the occurrences in Leicestershire, Yorkshire and Durham are new localities on the shores of reservoirs to which seed may have been carried naturally by birds.

9/15 GB post 1950 1 NNR : 2 SSSI TN = 7 R

Juncus capitatus Weigel Dwarf Rush

This dwarf annual of damp heaths, especially where water stands over-winter, is now known from only five localities in Cornwall, and from the Channel Isles where it still occurs in all the main islands. Formerly also reported from the Isles of Scilly and from Anglesey. Where decline has occurred it can only be due to loss of its particularly vulnerable habitat.

5/10 GB post 1950 1 NNR : 1 NCTR : 1 SSSI TN = 7 R
5/5 S post 1950

Juncus nodulosus Wahlenb. Marshall's Rush

A rare perennial known only from the stony shores of lochs in Aberdeen-shire and in Ross. It was reported to occur in two large colonies at the Ross locality in 1966 and two further, smaller, colonies were found in 1967. The Aberdeenshire record has not been confirmed recently.

2/2 GB post 1950 1 SSSI TN = 7 R

Juncus mutabilis Lam.   Pigmy Rush

A dwarf annual of damp hollows and wheel-ruts on heaths confined to the Lizard Peninsula of Cornwall. Population sizes vary with the season: over 13,000 plants were estimated to occur in one year, but two years later, in a dry spring, the total counted was only 172. In one station it is often very abundant locally in old tracks which are rather deep, but in another where the tracks are shallower, the species is rare due to competition from surrounding vegetation. Frequent disturbance is essential if this species is to survive. It flowers so early and is so inconspicuous that it is not threatened by collectors.

2/3 GB post 1960   1 NCTR : 3 SSSI   TN = 8 R

Juncus subulatus Forsk.

This tall perennial is confined to one locality in a saltmarsh in Somerset where it was first found in 1957. Two colonies are known and the plant appears to be well established. Almost certainly a recent arrival as the saltmarsh in which it occurs did not exist before 1910. Probably brought in by shipping or birds from the Mediterranean, its main centre of distribution.

1/1 GB post 1970   1 SSSI   TN = 9 R

Luzula pallescens Sw.   Fen Wood-rush

This tufted perennial is confined to open grassy places on peat in two localities in Huntingdonshire where fewer than 200 plants were recorded recently. In 1970 a small colony was found on the shores of Lough Neagh in Northern Ireland, but could not subsequently be refound. Other records from Surrey and Lincolnshire were either errors or introductions.

2/2 GB post 1970   2 NNR   TN = 8 R
1/1   H post 1970

Allium ampeloprasum L.   Wild Leek

This perennial of rocky and waste places near the coast is now apparently confined to Cornwall, Dorset and to Flatholme in the Bristol Channel. In Guernsey in the Channel Isles, where the plant has bulbils, it is frequent and spreading: it still occurs on Herm, but was only twice found in Jersey, in 1878 and 1924. Formerly apparently more widespread in Cornwall but it is easily confused with *Allium babingtonii* Borrer, and may well have been over-recorded. Other records from Steepholme (Somerset) and from Pembrokeshire have not recently been confirmed.

3/9 GB post 1950   No conservation   TN = 10 V
2/3   S post 1950

Allium babingtonii

Allium babingtonii Borrer   Babington's Leek

This perennial of rocks, sandy and waste places near the sea is now found
only in Cornwall and the Isles of Scilly. In Cornwall, although the plant is
widespread, the populations are generally small, rarely exceeding 20 plants.
In Scilly it has become extremely abundant since first recorded in 1939. It
is also widespread along the north and west coasts of Ireland, though recent
records are few. Other records from Devon and Dorset have not been
recently confirmed, and it is likely that the species was only a casual in
these counties.

12/16 GB post 1950   6 SSSI   TN = 5 R
 4/18 H post 1950

Allium sphaerocephalon L.   Round-headed Leek

The only mainland locality for this perennial is on limestone rocks in the
Avon Gorge near Bristol, where the populations are exposed to heavy
public pressure. Two separate colonies are known: one has only about ten
plants which are protected by their inaccessibility; the other is large but
more easily damaged by collectors. The species also still just survives in
Jersey in the Channel Isles.

1/1 GB post 1970   1 NNR   TN = 9 V
1/1 S post 1970

Leucojum vernum L.   Spring Snowflake

As a possible native this bulbous perennial only occurs in two localities in
Somerset and Dorset, where it grows in damp scrub and on stream-banks.
In Somerset it is abundant and well-protected on private land, whilst in
Dorset the population in the 1950s numbered over 1000 plants. However
it is regularly picked by local inhabitants and it has been reduced recently.
This species also occurs occasionally as a garden relic.

2/2 GB post 1950   No conservation   TN = 12 V

Leucojum aestivum L.   Summer Snowflake

This bulbous herb occurs in wet meadows and in willow thickets by rivers
in well over 30 localities in Wiltshire and Dorset, and in the valley of the
River Thames in Berkshire, Oxfordshire and Buckinghamshire. Formerly
also in Devon, Hampshire, Kent and Middlesex. It still occurs in at least
one locality in Limerick in western Ireland. Records from elsewhere are
undoubtedly of garden origin. Although losses in the past of this attractive

plant have been due to uprooting many of the present populations are large and protected by their relative inaccessibility.

14/29 GB post 1950   No conservation   TN = 8 R
1/4   H post 1970

Narcissus obvallaris Salisb.   Tenby Daffodil

This bulbous perennial is not known outside Britain and may be of garden origin. Although formerly recorded from pastures round Tenby, Pembrokeshire, and from Shropshire and Carmarthenshire, no truly wild populations are now known. The majority of plants which remain are maintained in gardens in the region in which it may have originated.

7/9 GB post 1950   2 SSSI   TN = 9 R

Iris spuria L.   Blue Iris

This perennial herb which may originally have been introduced into this country, still survives beside ditches and in old grassland in Dorset and in Lincolnshire. Of seven localities once known on Lincolnshire only one now remains. About 100 plants were recorded in Dorset in the 1960s.

2/2 GB post 1960   No conservation   TN = 12 V

Iris versicolor L.   Purple Iris

Although this attractive perennial was originally introduced it is well naturalised in marshy areas at the edge of lakes, ponds and streams in about eight localities in Essex, Westmorland, Kirkcudbrightshire, Inverness-shire, and the Outer Hebrides. Other records from West Yorkshire and Perthshire have not been recently confirmed.

6/10 GB post 1950   1 NCTR   TN = 9 R

Crocus purpureus Weston   Spring Crocus

Always introduced but occasionally naturalised and sometimes occurring in thousands in old meadows, pastures and churchyards. Known to have been established in Dorset, Isle of Wight, Hampshire, Berkshire, Oxfordshire, Norfolk, Gloucestershire and Nottinghamshire, and in one place in south-east Ireland. Though some sites in old meadows have been lost through cultivation, no general decline has been noted and at one Berkshire locality over 10,000 flowers were seen in 1976.

8/11 GB post 1950   1 NNR   TN = 8 R
1/1   H post 1930

Romulea columnae

Romulea columnae Seb. & Mauri   Sand Crocus

This small bulbous herb is known from only one locality in Devon where it grows in short sandy turf near the sea. The population is extensive spreading for over a mile on a golf course: its survival depending on the turf being kept short by mowing or grazing. It is common and widespread in the Channel Isles, and was formerly also in Cornwall where, however, it became extinct about 1880. The reason for the disappearance is uncertain but there are specimens in various herbaria collected over many years.

1/2 GB post 1970   1 NCTR: 1 SSSI   TN = 10 V
5/5 S post 1970

Gladiolus illyricus Koch   Wild Gladiolus

This bulbous perennial grows amongst bracken on bushy heaths, only in Hampshire now, though it was formerly on the Isle of Wight where it became extinct in 1897. About 40 populations are known, some of which may, however, have been planted. Heavy picking takes place especially where the species grows near car parks, but a recent field study has revealed a wider distribution than was previously known. As an introduction it has also been recorded from Devon and the Channel Isles.

7/9 GB post 1970   3 SSSI   TN = 7 R
                         Scheduled species

Cypripedium calceolus L.   Lady's-slipper

This most beautiful orchid is now at only one station in Yorkshire. It was formerly widespread, though local, in woods on Carboniferous, Magnesian and Corallian limestone in parts of the North Pennine region in Derbyshire, Yorkshire, Durham, Westmorland and Cumberland. Its virtual extinction has been due to uprooting and picking by gardeners, botanists and others, from very early times.

1/19 GB post 1970   1 SSSI   TN = 12 E
                         Scheduled species   V in Europe

Cephalanthera rubra (L.) Rich.   Red Helleborine

This attractive orchid of beechwoods on calcareous soils has been recorded, since 1960, in only the Chilterns in Buckinghamshire and in the Cotswolds in Gloucestershire. Formerly it may have been more widespread: there were other records from the Cotswolds and reports of its occurrence in Somerset, Hampshire, Sussex and Kent, though some of these were probably errors. The species was never very abundant anywhere, though colonies of up to 75 plants have been seen. Normally only a small percentage of plants produce flowers, and this species may well be overlooked.

3/9 GB post 1970	1 NNR Scheduled species	TN = 11 V

Epipactis dunensis (T. & T.A.Stephenson) Godfery   Dune Helleborine

This small, dull, endemic orchid occurs in 11 localities on stabilised sand dunes in Anglesey, Lancashire and Northumberland. Populations are, on the whole, small and, especially in south Lancashire, subject to severe public pressure. However it is locally common in one National Nature Reserve.

8/8 GB post 1960	2 NNR : 1 LNR	TN = 8 R

Epipogium aphyllum Sw.   Ghost Orchid

This saprophytic orchid occurs in deep shade in oak and beechwoods in Oxfordshire and Buckinghamshire; formerly it was also recorded from oakwoods in Herefordshire and Shropshire where it was last seen in 1910. It has always been very rare, usually with long intervals between successive flowering at any one site, though it has flowered almost annually at one of its sites in recent years. Normally only one or two plants appear and seed is rarely set. Even these plants are liable to attack by slugs and the rare-plant collector.

2/5 GB post 1970	2 SSSI Scheduled species	TN = 10 V

Spiranthes aestivalis (Poir.) Rich.   Summer Lady's-tresses

This orchid of bogs was once known from the New Forest in Hampshire and from the Channel Isles of Jersey and Guernsey, but was last seen in Hampshire in 1959 and must now be presumed extinct. It was first recorded in Hampshire in 1840 where, at one time, it occurred in quantity in four localities. By 1930 it had diminished considerably, mainly because of drainage, but almost certainly because of collecting as well. In Guernsey it was last seen in 1914 and was exterminated by a combination of drainage and disgraceful over-collecting. A similar fate befell the Jersey population where the last four plants, complete with tubers, were dug up in 1926.

0/1 GB post 1960	**EXTINCT**
0/2 S post 1926	

Spiranthes romanzoffiana

Spiranthes romanzoffiana Cham.   Irish Lady's-tresses
A species of moist meadows and pastures which are liable to flood in
winter. It also occurs in drained peaty areas, in upland bogs and on lake
margins. It has a very disjunct distribution: south-west, west and north-
east Ireland, Devon, Argyllshire and the Inner and Outer Hebrides. It is
relatively abundant in some Irish localities with over 100 plants seen in
some years, but numbers of flowering spikes fluctuate annually. Some
populations have been destroyed by drainage and reclamation. Cattle
grazing and trampling may also be detrimental though a small colony is
known to have persisted for nearly 20 years under these conditions in
Devon. In Argyllshire the populations fluctuate in time and space—it has
disappeared from several localities discovered in the 1960s but has now
been found in others. It still occurs in several of its localities in the Hebrides.

| 10/12 GB post 1950 | 2 SSSI | TN = 5 R |
| 11/19 H post 1930 | | R in Europe |

Hammarbya paludosa (L.) Kuntze   Bog Orchid
This small green orchid still occurs in numerous localities in bogs in the
north and west of Britain and in a few places in Ireland. However in south
and east England it is a declining species: many suitable habitats have been
destroyed by drainage. Although not yet threatened in the British Isles it
is under threat in Europe as a whole.   IUCN List, V in Europe

---

Liparis loeselii (L.) Rich.   Fen Orchid
This is one of our most rapidly declining orchid species. Once widespread
in the fens of East Anglia and in dune slacks in South Wales and known
from over 30 localities it is now known from only eight in Devon, Norfolk,
Glamorgan and Carmarthenshire, and several of these are represented by
single figure populations. The decline of this species has been almost
entirely due to drainage and it seems very unlikely that it can survive much
longer in East Anglia, where even nature reserves give no protection
against the general over-drainage of the region.

| 8/29 GB post 1960 | 3 NNR : 3 SSSI | TN = 9 V |
| | | V in Europe |

Neotinea maculata (Desf.) Stearn   Dense-flowered Orchid
N. intacta (Link) Reichb.f.
This Mediterranean orchid was first found in Great Britain in the Isle of
Man in 1967, where a small population of up to 20 spikes now occurs
regularly protected only by its early flowering. Otherwise the species is
confined to the western half of Ireland where it is characteristic of the

limestone country, occurring in a great range of habitats and appearing capable of withstanding even heavy grazing. Off the limestone it is found on calcareous sand, as in Man, and even, in a few places, on light peat overlying acid rocks. Though the species appears to have declined in Ireland since 1930, this may be due solely to under-recording of this rather inconspicuous species. Indeed its recent discovery in East Cork and in the Aran Isles suggests it may still be spreading.

1/1   GB post 1970   No conservation   TN = 9 R
19/32   H post 1930

Ophrys fuciflora (Crantz) Moench   Late Spider-orchid

This orchid occurs on chalk downs and in field borders in Kent where about ten localities are known, in several of which it is plentiful, though in others limited to one of two specimens. In the past it was apparently much more common and generally distributed in Kent, but destruction of downland and undergrazing in the last half century have brought about a heavy decline. Other records from Dorset, Surrey, Suffolk and Gloucestershire were almost certainly errors.

4/6 GB post 1960   4 NNR : 5 SSSI   TN = 9 R

Ophrys sphegodes Mill.   Early Spider-orchid

This tuberous orchid of dry banks and grassy slopes on chalk or limestone is now known from only 13 localities in England south of the River Thames in Dorset, Hampshire, Sussex, Kent and Gloucestershire. Formerly much more widespread and recorded from Cornwall, Wiltshire, Isle of Wight, Surrey, Essex, Oxfordshire, Suffolk, Cambridgeshire, Bedfordshire, Northamptonshire, Denbighshire and from Jersey in the Channel Isles. However in the counties north of the River Thames nearly all these records date from before 1850 and the major decline of this species took place long before the ploughing up of old chalk and limestone grassland. Nevertheless ploughing and public pressure are now having a deleterious effect, especially on some Dorset populations.

10/53 GB post 1960   1 NNR : 8 SSSI   TN = 10 V
 0/1   S post 1929

Ophrys bertolonii Mor.

This orchid was recorded for the first time in Great Britain in 1976 when a small colony was discovered in Dorset. It was in an area frequently visited by botanists and it seems likely it was a recent introduction, though possibly by natural means.

1/1 GB post 1970   No conservation   TN = 12 V

Himantoglossum hircinum

Himantoglossum hircinum (L.) Spreng.   Lizard Orchid

This handsome orchid occurs in wood margins, by woodland paths, in scrub, on field borders and in open grassland, invariably on calcareous soils — chalk, limestone or permanent dunes. Records have been made since 1960 in Devon, Wiltshire, Sussex, Kent, Suffolk, Cambridgeshire and Jersey in the Channel Isles, though it has since disappeared from the last named. Populations of this species have fluctuated constantly: thought to be almost extinct in about 1900 there was a rapid expansion between 1920 and 1940. Since 1940 a new decline has begun and now only a very few permanent colonies survive.

| 10/98 GB post 1960 | 6 SSSI | TN = 10 V |
| 0/1   S post 1970 | | |

Orchis militaris L.   Military Orchid

A handsome orchid of beech and other woodland, scrub and wood borders on chalk, occurring now only in Buckinghamshire, where numbers are small, and in Suffolk where there is a large population. It was formerly also on Hertfordshire, Middlesex, Berkshire and Oxfordshire and there are unconfirmed records for Surrey. Over the last two centuries it has been recorded from a number of localities scattered over the Chiltern Hills but it was probably never in more than a few of them at any one time, and was over collected. Because of changes in the management of its native beech woodland, it had been believed extinct in Buckinghamshire until refound in 1947, where the population has one year since exceeded 100 but is now much smaller. It is very susceptible, especially in the young stages, to damage from trampling e.g. by photographers, so this site is now enclosed. The Suffolk population is similarly protected. Natural changes in the habitat also present difficult management problems.

| 2/19 GB post 1970 | 2 NCTR | TN = 11 V |
| | Scheduled species | |

Orchis simia Lam.   Monkey Orchid

This attractive orchid is found on base-rich grassland, in bushy places and on field borders. It is now known from two localities in Kent (one a transplant to conserve the species), from two in Oxfordshire, and from a recently discovered site in Yorkshire. Formerly it also occurred in Sussex, Surrey, Berkshire and Buckinghamshire. Many colonies have been destroyed by removal and burning of turf, by over-collecting, rabbit grazing and ploughing. The best part of one Oxfordshire locality was lost through ploughing about 1950 and the few remaining plants which flower annually are often picked despite wardening by the local Nature Conservation Trust.

3/8 GB post 1970   2 NCTR : 1 SSSI   TN = 11 V
                  Scheduled species

Eriophorum gracile Roth   Slender Cottongrass

This creeping perennial of wet acid bogs has been recorded since 1960 from Hampshire, Surrey and Anglesey. It seems to have disappeared from Dorset and Norfolk in the 1950s and it is almost certainly extinct in Somerset, Northamptonshire and Yorkshire. This is a very difficult species to conserve: it requires extremely wet conditions and is therefore vulnerable to drainage and reclamation. It was not discovered in Ireland until 1966 where it must have been previously overlooked as the sites are undoubtedly native.

5/13 GB post 1960   2 SSSI   TN = 8 V
6/6   H post 1960

Scirpus hudsonianus (Michx.) Fernald

This creeping perennial was first recorded in its sole British station, a bog in Angus, in 1791 but became extinct when the habitat was dredged for marl and subsequently flooded, about 1813. Other records for this species were apparently errors.

0/1 GB post 1813   **EXTINCT**

Scirpus holoschoenus L.   Round-headed Club-rush

This tufted perennial occurs as a native on damp sandy flats by the sea in Devon and Somerset. The Devon population is large and well protected, but the Somerset population is very small. Also reported, almost certainly as an introduction, from Kent, Monmouthshire, Glamorgan and from Jersey in the Channel Isles. There are eighteenth century records from Hampshire of uncertain status.

2/2 GB post 1970   1 NNR : 1 SSSI   TN = 6 R

Scirpus triquetrus L.   Triangular Club-rush

This creeping perennial of muddy banks of tidal rivers now occurs only on the River Tamar on the borders of Cornwall and Devon where several small populations have been recorded recently. Elsewhere, in Sussex, Essex, Kent, Surrey and Middlesex, where it occurred mainly by the River Thames and its tributaries, it is now apparently extinct, as a result of embanking. It still flourishes in Ireland where it is frequent for about five miles along the Shannon estuary and its tributaries.

1/8 GB post 1970   No conservation   TN = 10 E
2/3   H post 1950

Eleocharis parvula

Eleocharis parvula (Roem. & Schult.) Link   Dwarf Spike-rush

This slender perennial of wet muddy estuarine shores has been recorded since 1950 from Devon, Dorset, Hampshire, Merionethshire and Caernarvonshire. Some of the populations are extensive and, though now apparently absent from some former localities, it has a history of being 'refound' after long time intervals and it is clear that this insignificant species is frequently overlooked. However it does appear to be threatened by gold dredging in North Wales. Also recorded in Ireland but none of it sthree localities has been confirmed for 25 years.

7/10 GB post 1950   2 NNR   TN = 4 R
3/3   H post 1930

Eleocharis austriaca Hayek   Northern Spike-rush

This perennial rush forms vigorous stands on sandy ground by upland rivers in Yorkshire, Northumberland, Cumberland and Selkirk where 13 populations are known. It was not discovered in Britain until 1947 when material was collected in Yorkshire and not identified until 1960. It is easily overlooked and may well occur in other parts of the Borders.

10/10 GB post 1960   No conservation   TN = 4 R

Cyperus fuscus L.   Brown Galingale

This annual species occurs in damp places especially on bare ground left by the drying-up of ponds and ditches. It has been recorded since 1960 only from Somerset, Hampshire and Middlesex. It was last seen in Jersey in the Channel Isles in 1928 and was also formerly in Dorset, Surrey, Berkshire and Buckinghamshire. Only one of the three existing populations is of any size, and here the numbers are erratic both in occurrence and quantity from year to year. The serious decline of this species has almost certainly been caused by elimination of the habitat.

3/9 GB post 1970   No conservation   TN = 10 E
0/2   S post 1950

Schoenus ferrugineus L.   Brown Bog-rush

This tufted glabrous perennial once grew luxuriently in base-rich flushes at the wet, peaty margin of Loch Tummel in Perthshire but became extinct in 1953 following the raising of the water-level for a hydro-electric scheme. At that time plants were transferred to the new margin of the loch, where they failed to become established, and to a mountain site in Perthshire where, although the plants survive and appear healthy, very little spread has occurred. In 1975 it was reintroduced onto the shores of Loch Tummel where it is hoped the habitat is now more suitable.

0/1 GB post 1953   **EXTINCT**

Kobresia simpliciuscula (Wahlenb.) Mackenzie   False Sedge

This tufted perennial of stony and grassy flushes occurs in the mountainous regions of Yorkshire, Durham and Perthshire. It may also still persist in Westmorland and Argyllshire but no recent records exist. It occurs in great quantity in Teesdale but is not so frequent on the Yorkshire side where it is more heavily grazed. In Perthshire it is local and most of the populations are small.

10/14 GB post 1950   2 NNR : 1 SSSI   TN = 3 R

Carex flava L.   Large Yellow-sedge

This very critical sedge occurs on damp peaty soils overlying Carboniferous limestone at one locality in Lancashire. The population previously recorded from Yorkshire consists of hybrids with *C. lepidocarpa* Tausch, and a population consisting of similar intermediates has been recorded from Galway in Western Ireland.

1/1 GB post 1970   1 NNR   TN = 6 R

Carex depauperata Curt.   Starved Wood-sedge

This creeping sedge of dry woods and hedge banks on chalk and limestone is one of our most endangered species. It is now found only in Somerset where as few as eight plants have a precarious existence in a lane bank. The species has disappeared since 1972 from a similar habitat in Surrey where one plant persisted perilously for many years. Formerly also apparently in Anglesey where it was last seen in 1936. The only other population in the British Isles is that discovered in Cork in southern Ireland in 1973.

1/6 GB post 1970   1 SSSI   TN = 12 E
1/1 H post 1970

Carex filiformis L.   Downy-fruited Sedge

This shortly creeping sedge occurs in damp meadows, limestone grassland, dry roadsides, and in woodland rides. About ten stations are known in Wiltshire, Surrey, Oxfordshire and Gloucestershire. It was recorded in Middlesex until 1960 but the area has been disturbed by gravel digging and it is now probably extinct. The species is rare in the majority of its localities.

9/13 GB post 1960   2 NCTR : 2 SSSI   TN = 6 R

Carex rariflora

Carex rariflora (Wahlenb.) Sm.   Mountain Bog-sedge

This creeping perennial has recently been recorded from only seven
localities on slopes of oligotrophic peat at about 2500 ft. (800 m.) in
Perthshire, Angus, Aberdeenshire and Inverness-shire. It was formerly also
reported from Banff. This apparent decline is almost certainly due to
under-recording: the plant is plentiful in several localities and in no danger.
7/12 GB post 1950   3 NNR : 2 SSSI   TN = 4 R

Carex ornithopoda Willd.   Bird's-foot Sedge

This tufted perennial grows on well-drained dry limestone grassland or in
crevices in limestone pavement. About 15 colonies have been recorded
since 1950 in Derbyshire, Yorkshire and Westmorland but an earlier
record from Cumberland has not been refound. Several of the extant
colonies are extensive with too many plants to be counted accurately.
Although there has been a slight decrease the present situation is fairly
satisfactory as several sites have some measure of protection.
10/11 GB post 1950   1 NCTR : 4 SSSI   TN = 4 R

Carex buxbaumii Wahlenb.   Club Sedge

A creeping perennial of mesotrophic fens now to be found at only two
stations in Inverness-shire, in one of which it is locally plentiful: in the
other it is much reduced and in danger from over-grazing by sheep. Also
formerly occurred on a small island in Lough Neagh, Northern Ireland, but
became extinct there about 1886 through over-grazing and trampling by
cattle.
2/2 GB post 1960   1 SSSI   TN = 8 V
0/1   H post 1886

Carex norvegica Retz.   Close-headed Alpine-sedge

This tufted perennial grows in small quantity on wet ledges and on rocky
slopes in the higher mountains of Perthshire, Angus and Aberdeenshire.
This is a rare sedge which was over-collected in the past and this may be the
reason that it has now possibly gone from one Perthshire station where it
has not been recorded since 1938.
3/4 GB post 1950   1 NNR : 2 SSSI   TN = 5 R

Carex atrofusca Schkuhr.   Scorched Alpine-sedge

This creeping sedge grows in micaceous stony flushes between 1800 and
3500 ft. (500-1070 m.). in the mountains of Perthshire and Inverness-shire

where four colonies are known. In Perthshire the species is plentiful in one locality, whilst 50-100 plants occur in another: in the sole station in Inverness-shire it occurs in fair quantity in a remote corrie. A threat exists from collectors, particularly at the one very well-known locality.

4/5 GB post 1960   1 NT : 3 SSSI   TN = 4 R

Carex recta Boott   Estuarine Sedge

This shortly creeping perennial grows in the lower reaches of rivers in stiff, peaty alluvium, usually where silt is periodically deposited and where there is a seasonally fluctuating water table. It is now known from at least six localities in Inverness-shire, Ross, Sutherland and Caithness in two of which the species forms very strong stands and is dominant over small areas.

3/4 GB post 1970   No conservation   TN = 6 R
                                                       R in Europe

Carex chordorrhiza L.f.   String Sedge

This extensively creeping perennial is known from three localities within a restricted area in Sutherland where it grows in very wet, base-poor bogs. Although thus restricted one of the colonies consists of over 10,000 spikes and the species is in no apparent danger.

1/1 GB post 1970   1 SSSI   TN = 6 R

Carex lachenalii Schkuhr   Hare's-foot Sedge

This creeping perennial occurs in high-level flushes and on rocky ledges in Angus, Aberdeenshire, Banffshire, Inverness-shire and Argyllshire where at least seven colonies are now known. It is limited by the extent of its habitat — extremely late snow beds — and, though locally abundant at one station in Banffshire, it is sparse and local in most of the others, and is subject to much public pressure in at least one locality.

6/7 GB post 1950   2 NNR : 4 SSSI   TN = 4 R

Carex microglochin Wahlenb.   Bristle Sedge

The only station for this creeping perennial is in the mountains of Perthshire where, in 1970, it was abundant over two kilometre squares. It grows at over 2500 ft. (800 m.) in gently sloping, stony, micaceous flushes where the total plant cover is usually less than 50 per cent.

1/1 GB post 1970   1 NNR   TN = 4 R

Carex davalliana

Carex davalliana Sm.

This tufted perennial formerly occurred in a boggy place near Bath in Somerset but became extinct about 1845 when the site was drained or built upon. It was also recorded from two places in Yorkshire where it was of uncertain status.

0/3 GB post 1845   **EXTINCT**

Leersia oryzoides (L.) Sw.   Cut-grass

This erect perennial of wet meadows, canal and river-sides and ditch margins has been recorded from only ten localities in Somerset, Dorset, Sussex and Surrey since 1960 and it has subsequently disappeared from at least two of these. It was formerly also in Hampshire but there are no records for over 30 years.

7/21 GB post 1960   1 SSSI   TN = 9 V

Festuca caesia Sm.   Blue Fescue

This tufted grass has been recorded from a few scattered areas in dry acid grassland and on cliffs in Devon, Suffolk, Anglesey, Lincolnshire and the Channel Isles. It appears to be native in most of these and may well have been overlooked in the past.

4/4 GB post 1930   1 LNR : 1 SSSI   TN = 5 R
2/2   S post 1960

Poa infirma Kunth   Early Meadow-grass

This upright annual grows in sandy places usually near the sea in the Isles of Scilly, in Cornwall and in the Channel Isles. It flowers in March and April and it is difficult to find after the middle of the latter month. Perhaps for this reason it was not discovered in the Channel Isles until 1914 or in Scilly and Cornwall until 1950. This early disappearance doubtless helps to protect the species from the attentions of collectors. It is often abundant, especially on footpaths, bare ground and disturbed areas and is apparently in no danger.

5/5 GB post 1950   1 NNR : 6 SSSI   TN = 5 R
5/5   S post 1970

Poa flexuosa Sm.   Wavy Meadow-grass

This tufted perennial of mountain screes and ledges is found in eight localities in Aberdeenshire, Inverness-shire and Ross between 2500 and 3600 ft. (800-1100 m.). In at least two stations in Aberdeenshire and Inver-

76

ness-shire small populations are threatened by public pressure, including climbing, whilst in Ross tillers, which are broken off from tussocks by deer trampling, subsequently die.

7/7 GB post 1950   1 NNR : 2 SSSI   TN = 3 R

Bromus madritensis L.   Compact Brome

This erect annual grass of dry, rather open habitats is apparently native on sand and limestone soils in Somerset, Gloucestershire, Pembrokeshire and the Channel Isles, but it is also widespread as an alien and persists for many years on old walls and in similar habitats. It has also been recorded from south-east Ireland and persists in one locality in Waterford.

1/4 GB post 1970   1 NNR : 1 SSSI   TN = 8 R
4/4   S post 1970
1/2   H post 1950

Bromus tectorum L.   Drooping Brome

This tufted annual is naturalised and locally abundant on sandy tracks, on firebreaks and in gravel pits in the Breckland of Suffolk, Norfolk and Cambridgeshire where at least 12 colonies are now known. It is possible that the species is still spreading here. Elsewhere it has been widely recorded as a casual.

5/5 GB post 1960   No conservation   TN = 7 R

Bromus interruptus (Hack.) Druce   Interrupted Brome

This endemic annual grass of arable fields is now almost certainly extinct: it has not been seen in its last known station in Cambridgeshire since 1972. It probably arose as a mutant in a sainfoin field in the same county about 1870. It became widespread in south-east England and by the 1920s had been recorded from about 65 ten kilometre squares. Subsequent decline was probably due to the use of better seed-cleaning methods. By 1962 only one site was known. Conservation action by the local Nature Conservation Trust did not succeed in saving the species.

0/65 GB post 1972   **EXTINCT**

Agropyron donianum F.B.White   Don's Couch

Until 1951 only one locality for this tufted perennial of limestone rocks was known, but 16 colonies have now been reported from Perthshire, Aberdeenshire, Banffshire, Sutherland and Caithness. Doubtless further sites still remain undiscovered and the species is in no way threatened.

11/11 GB post 1960   5 NNR : 3 SSSI   TN = 3 R

Koeleria vallesiana

Koeleria vallesiana (Honck.) Bertol.   Somerset Hair-grass

This densely tufted perennial is known only from Somerset where, however, it is locally abundant in short grassy turf or on ledges on rocky limestone slopes. At least 12 localities have been reported from six major sites at some of which it occurs in thousands. The species is in no danger as most of these localities are adequately protected.

4/4 GB post 1960   7 SSSI   TN = 5 R

Calamagrostis scotica (Druce) Druce   Scottish Small-reed

This endemic grass is confined to marshes and bogs in Roxburghshire and Caithness where only two localities are now known. Formerly also recorded from two other localities in Caithness. Although the existing Caithness population is scattered over 100 m. it might be flooded if plans to build an oil-drilling platform on the coast nearby were put into effect.

2/4 GB post 1960   No conservation   TN = 9 V

Mibora minima (L.) Desv.   Early Sand-grass

This small tufted annual has been recorded as a native in sandy places near the sea only from Glamorgan and Anglesey, where four stations are known, and the Channel Isles where it is locally plentiful: it is found as an established introduction in Dorset, Suffolk, Bedfordshire and East Lothian. Also recorded as a casual from Cornwall, Hampshire, Surrey, Cambridgeshire and Northumberland. In one station in Anglesey, part of the habitat has suffered through having been ploughed by the Forestry Commission.

4/4 GB post 1950   2 NNR   TN = 7 R
3/4   S post 1970

Gastridium ventricosum (Gouan) Schinz & Thell.   Nit-grass

This tufted annual appears to be native in dry, open sandy or calcareous grassland in southern Britain, mainly near the sea. At one time it was recorded from about 28 vice-counties extending as far north as Yorkshire, though it was undoubtedly only casual in many inland stations. However, since 1960 it has only been reported from six localities in Cornwall, Somerset, Dorset, Hampshire, Gloucestershire and Glamorgan, and from Guernsey and Sark in the Channel Isles.

6/125 GB post 1960   2 NT   TN = 9 V
2/2   S post 1970

Phleum phleoides (L.) Karst.   Purple-stem Cat's-tail

This erect perennial is known from over 30 localities in dry sandy and chalky pastures in the Breckland of Suffolk and Norfolk, with outlying localities on the borders of Hertfordshire and Bedfordshire, and in Cambridgeshire. It is locally abundant in some parts of Breckland where its ability to recolonise bare ground on roadside verges, tracks and in gravel pits, coupled with its tolerance of rabbit-grazing, ensures its survival. However many of the outlying localities have been lost or are threatened: in Hertfordshire recently only three plants were observed though in the past there had been many more, a reduction due to erosion of the habitat by children playing there. It was last seen in its sole Essex locality in 1861.

10/22 GB post 1960   2 NNR : 1 NCTR : 4 SSSI   TN = 6 R

Alopecurus bulbosus Gouan   Bulbous Foxtail

This slender tufted perennial of grassy salt-marshes has been recorded from 25 vice-counties in England and Wales as far north as the estuaries of the Humber and the Mersea. However it now appears to be extinct in many of its former northern localities and all post 1960 records have been on or south of a line joining the Thames and the Severn estuaries, in the counties of Devon, Dorset, Hampshire, Sussex, Kent, Gloucestershire, Monmouthshire and Carmarthenshire. It still persists in a single locality in the Channel Isles, on Guernsey, where it was first discovered in 1958. No cause for this decline is known but this may be yet another maritime species, with its centre of distribution in the Mediterranean, reaching its northern limit in Britain, which has been affected by a change in our climate.

10/60 GB post 1960   No conservation   TN = 10 V
 1/1   S post 1960

Hierochloe odorata (L.) Beauv.   Holy-grass

This early-flowering tufted perennial of wet meadows, grassland by the sea and river banks has been recorded from 17 localities in Kirkcudbrightshire, Renfrewshire, Roxburghshire, Fife, Kinross, Caithness and the Outer Hebrides. Many of these are recent discoveries and it is clear this species has been much under-recorded in the past. However it is extinct in Angus, where it was first discovered in Britain in 1821. In Ireland it still occurs on the shores of Lough Neagh in Antrim.

11/12 GB post 1960   3 NNR : 4 SSSI   TN = 3 R
 1/1   H post 1950

Anthoxanthum puelii

Anthoxanthum puelii Lecoq & Lamotte   Annual Vernal-grass

This annual weed of cultivated and waste land was probably introduced from France in the latter half of the nineteenth century with seeds of fodder plants. It is occasionally persistent particularly in southern England where it has been recorded recently from Surrey, Suffolk and Cambridgeshire. However it may already be extinct in both the last two whilst the Surrey populations are small and unreliable in appearance. Formerly widespread as a casual but much less common now probably as a result of the use of purer seed in agriculture.

6/65 GB post 1960   No conservation   TN = 10 E
0/2   S post 1897

Spartina alterniflora Lois.   Smooth Cord-grass

This perennial, which was introduced accidentally from North America before 1836, has persisted ever since in Southampton Water. Though formerly more widespread one large population remains in the upper reaches of the estuary.

1/4 GB post 1970   No conservation   TN = 10 V

Cynodon dactylon (L.) Pers.   Bermuda-grass

This extensively creeping perennial is probably native and is certainly still persistent in Cornwall, Somerset, Dorset, Hampshire, Essex and the Channel Isles, particularly near the coast. It is also widespread as an alien, usually only casual, but occasionally persistent for some years, as for example in the London area. Many of the sites are close to human habitation and are therefore subject to public pressures.

10/12 GB post 1950   2 SSSI   TN = 6 R
 1/1   S post 1970

APPENDIX I

Taxa considered for inclusion but discarded as they now occur in over 15 10 kilometre squares

Allium schoenoprasum

Alopecurus alpinus

Briza minor

Callitriche truncata

Carex digitata

C.   montana

C.   rupestris

Circaea alpina

Corallorhiza trifida

Corynephorus canescens

Crepis mollis

Cystopteris montana

Dactylorhiza traunsteineri

Daphne mezereum

Draba norvegica

Elodea nuttallii

Festuca juncifolia

Gentianella germanica

Illecebrum verticillatum

Juncus alpinoarticulatus

Lathyrus palustris

Linnaea borealis

Linum perenne

Luzula arcuata

Marrubium vulgare

Melampyrum sylvaticum

Nardurus maritimus

Oenanthe silaifolia

Polypogon monspeliensis

Rhyncospora fusca

Ruppia spiralis

Salix arbuscula

S.   reticulata

Sonchus palustris

Stratiotes aloides

Tilia platyphyllos

Trifolium occidentale

Verbascum pulverulentum

V.   virgatum

Veronica alpina

V.   spicata subsp. *hybrida*

Vicia lutea

TABLE 1
Table of Threat Numbers (TN)
Columns marked *I* contribute to Threat Number
(for further explanation see pp. x-xiii)

SPECIES	1 Ireland (H)	2 Channel Isles (S)	3 Great Britain (GB) *I*	4 Total *I*	5 Attractiveness *I*	6 Conservation *I*	7 Remoteness *I*	8 Accessibility *I*	9 TN	10 IUCN Category
		Rate of decline		Local'es						

EXTINCT (20 species)

SPECIES	1	2	3	4	5	6	7	8	9	10
Ajuga genevensis	—	—	0/2							EX
Bromus interruptus	—	—	0/65							EX
Bupleurum falcatum	—	—	0/2							EX
Campanula persicifolia	—	—	0/5							EX
Carex davalliana	—	—	0/3							EX
Centaurium latifolium	—	—	0/1							EX
Euphorbia villosa	—	—	0/1							EX
Filago gallica	—	1/2	0/10							EX
Fumaria martinii	—	0/1	0/10							EX
Halimione pedunculata	—	—	0/14							EX
Holosteum umbellatum	—	—	0/5							EX
Hydrilla verticillata	1/1	—	0/1							EX
Otanthus maritimus	1/5	0/3	0/18							EX
Pinguicula alpina	—	—	0/1							EX
Rubus arcticus	—	—	0/2							EX
Saxifraga rosacea	18/19	—	0/1							EX
Schoenus ferrugineus	—	—	0/1							EX
Scirpus hudsonianus	—	—	0/1							EX
Senecio congestus	—	—	0/24							EX
Spiranthes aestivalis	—	0/2	0/1							EX

TN = 13 (8 species)

SPECIES	1	2	3		4	5	6	7	8	9	10
Agrostemma githago	2/59	0/4	4/∞	2	4 3	2	2	2	2	13	E
Alyssum alyssoides	—	—	2/6	2	2 4	1	2	2	2	13	E
Damasonium alisma	—	—	1/50	2	1 4	1	2	2	2	13	E
Galium spurium	—	—	1/4	2	1 4	1	2	2	2	13	E
Petrorhagia nanteuilii	—	1/1	1/4	2	2 4	1	2	2	2	13	E
Pyrus cordata	—	—	2/4	1	2 4	1	3	2	2	13	E
Senecio paludosus	—	—	1/8	2	1 4	1	2	2	2	13	V
Stachys germanica	—	—	2/10	2	5 3	2	2	2	2	13	E

SPECIES	1	2	3		4		5	6	7	8	9	10
			Rate of decline		Local'es		Attractiveness	Conservation	Remoteness	Accessibility	TN	IUCN Category
	Ireland (H)	Channel Isles (S)	Great Britain (GB) **I**		Total **I**		**I**	**I**	**I**	**I**		

TN = 12 (18 species)

SPECIES	1	2	3		4		5	6	7	8	9	10
Althaea hirsuta	—	—	1/2	1	2	4	2	1	2	2	12	E
Armeria maritima subsp. elongata	—	—	2/6	2	3	3	2	1	2	2	12	V
Campanula rapunculus	—	—	7/75	2	7	2	2	2	2	2	12	V
Carex depauperata	1/1	—	1/6	2	1	4	0	2	2	2	12	E
Caucalis platycarpos	—	—	3/81	2	3	3	1	2	2	2	12	E
Centaurium tenuiflorum	—	0/2	2/3	1	2	4	1	2	2	2	12	V
Cyclamen hederifolium	—	—	2/4	1	2	4	2	1	2	2	12	V
Cypripedium calceolus	—	—	1/19	2	1	4	2	1	1	2	12	E
Galeopsis segetum	—	—	1/24	2	1	4	1	2	1	2	12	E
Gnaphalium luteoalbum	—	2/2	1/5	2	1	4	0	2	2	2	12	E
Iris spuria	—	—	2/2	0	2	4	2	2	2	2	12	V
Leucojum vernum	—	—	2/2	0	2	4	2	2	2	2	12	V
Ophrys bertolonii	—	—	1/1	0	1	4	2	2	2	2	12	V
Orobanche caryophyllacea	—	—	2/4	1	2	4	2	1	2	2	12	E
Scheuchzeria palustris	0/1	—	2/9	2	2	4	1	2	1	2	12	E
Stachys alpina	—	—	2/3	1	2	4	2	1	2	2	12	E
Trifolium stellatum	—	—	1/1	0	1	4	1	3	2	2	12	E
Veronica triphyllos	—	—	1/24	2	1	4	0	2	2	2	12	E

SPECIES	1 Ireland (H)	2 Channel Isles (S)	3 Great Britain (GB)	I	4 Total	I	5 Attractiveness I	6 Conservation I	7 Remoteness I	8 Accessibility I	9 TN	10 IUCN Category
	1	2	3		4		5	6	7	8	9	10
			Rate of decline		Local'es							

TN = 11 (20 species)

SPECIES	1	2	3		4		5	6	7	8	9	10
Artemisia campestris	—	—	2/11	2	6	2	1	2	2	2	11	E
Campanula patula	—	—	14/91	2	15	1	2	2	2	2	11	V
Cephalanthera rubra	—	—	3/9	2	3	3	2	1	1	2	11	V
Cotoneaster integerrimus	—	—	1/1	0	1	4	2	2	2	1	11	E
Eryngium campestre	—	1/1	4/15	2	4	3	1	1	2	2	11	V
Geranium purpureum subsp. forsteri	—	0/1	3/7	1	3	3	1	2	2	2	11	V
Lactuca saligna	—	—	4/32	2	8	2	1	2	2	2	11	E
Lonicera xylosteum	—	—	1/1	0	1	4	1	2	2	2	11	V
Lythrum hyssopifolia	—	1/2	3/38	2	3	3	0	2	2	2	11	E
Matthiola sinuata	0/8	2/2	2/15	2	6	2	2	1	2	2	11	V
Melampyrum arvense	—	—	5/40	2	6	2	1	2	2	2	11	E
Orchis militaris	—	—	2/19	2	2	4	2	0	2	1	11	V
O.　　 simia	—	—	3/8	1	4	3	2	1	2	2	11	V
Orobanche loricata	—	0/1	5/23	2	6	2	1	2	2	2	11	V
O.　　 reticulata	—	—	4/6	1	4	3	1	2	2	2	11	E
Pulicaria vulgaris	—	0/3	9/117	2	9	2	1	2	2	2	11	V
Scleranthus perennis subsp. prostratus	—	—	2/11	2	3	3	0	2	2	2	11	E
Silene italica	—	—	1/1	0	1	4	1	2	2	2	11	V
Spergularia bocconii	—	2/3	2/9	2	2	4	0	2	1	2	11	E
Veronica spicata subsp. spicata	—	—	3/10	2	4	3	2	1	1	2	11	V

SPECIES	1	2	3		4		5	6	7	8	9	10
	Rate of decline				Local'es		Attractiveness	Conservation	Remoteness	Accessibility		IUCN Category
	Ireland (H)	Channel Isles (S)	Great Britain (GB)	I	Total	I	I	I	I	I	TN	

TN = 10 (37 species)

SPECIES	1	2	3		4		5	6	7	8	9	10
Allium ampeloprasum	—	2/3	3/9	2	3	3	1	2	1	1	10	V
Alopecurus bulbosus	—	1/1	10/60	2	15	1	1	2	2	2	10	V
Anthoxanthum puelii	—	0/2	6/65	2	9	2	0	2	2	2	10	E
Apium repens	—	—	1/2	1	1	4	0	1	2	2	10	E
Arenaria norvegica subsp. anglica	—	—	2/2	0	2	4	1	2	1	2	10	E
Arnoseris minima	—	—	4/76	2	6	2	0	2	2	2	10	E
Asparagus officinalis subsp. prostratus	5/6	2/2	4/12	2	5	3	1	1	2	1	10	V
Bupleurum baldense	—	4/4	2/3	1	2	4	0	1	2	2	10	V
B. rotundifolium	—	—	8/150	2	8	2	0	2	2	2	10	E
Cynoglossum germanicum	—	—	6/50	2	8	2	1	1	2	2	10	V
Cyperus fuscus	—	0/2	3/9	2	4	3	0	2	2	1	10	E
Cystopteris dickieana	—	—	1/3	2	1	4	1	1	1	1	10	E
Dianthus gratianopolitanus	—	—	3/3	0	5	3	2	2	2	1	10	V
Diapensia lapponica	—	—	1/1	0	1	4	2	2	0	2	10	V
Epipogium aphyllum	—	—	2/5	1	4	3	1	1	2	2	10	V
Gagea saxatilis	—	—	1/1	0	1	4	1	1	2	2	10	E
Galium tricornutum	—	—	13/∞	2	13	1	1	2	2	2	10	E
Geranium purpureum subsp. purpureum	1/2	3/3	8/22	1	9	2	1	2	2	2	10	V
Himantoglossum hircinum	—	0/1	10/98	2	11	1	2	1	2	2	10	V
Hypericum linarifolium	—	3/4	4/10	1	5	3	1	2	1	2	10	V
Limonium recurvum	—	—	1/1	0	1	4	1	2	2	1	10	E
Linaria supina	—	—	6/15	1	7	2	1	2	2	2	10	V
Mentha pulegium	0/32	1/3	14/∞	2	12	1	1	2	2	2	10	V
Ophrys sphegodes	—	0/1	10/53	2	13	1	2	1	2	2	10	V
Romulea columnae	—	5/5	1/2	1	2	4	1	1	1	2	10	V
Sagittaria rigida	—	—	2/2	0	2	4	1	1	2	2	10	V
Salvia pratensis	—	—	14/28	1	14	1	2	2	2	2	10	V
Scirpus triquetrus	2/3	—	1/8	2	3	3	0	2	2	1	10	E
Scleranthus perennis subsp. perennis	—	—	1/1	0	1	4	0	2	2	2	10	E
Scorzonera humilis	—	—	1/3	2	1	4	0	0	2	2	10	V
Sisymbrium irio	2/5	—	14/48	2	15	1	1	2	2	2	10	V
Sorbus subcuneata	—	—	2/3	1	3	3	1	2	1	2	10	V
S. vexans	—	—	2/2	0	2	4	1	2	1	2	10	V
Spartina alterniflora	—	—	1/4	2	1	4	0	2	1	1	10	V
Teucrium scordium	8/19	0/2	3/22	2	3	3	1	0	2	2	10	V
Tuberaria guttata subsp. breweri	4/7	—	4/5	0	5	3	2	1	2	2	10	V
Viola persicifolia	7/11	—	3/17	2	3	3	1	1	1	2	10	E

SPECIES	1	2	3		4		5	6	7	8	9	10
			Rate of decline		Local'es							
	Ireland (H)	Channel Isles (S)	Great Britain (GB) I		Total I		Attractiveness I	Conservation I	Remoteness I	Accessibility I	TN	IUCN Category

TN = 9 (53 species)

SPECIES	1	2	3		4		5	6	7	8	9	10
Alchemilla minima	—	—	2/2	0	2	4	1	1	1	2	9	V
A. subcrenata	—	—	2/2	0	5	3	1	2	1	2	9	V
Allium sphaerocephalon	—	1/1	1/1	0	2	4	1	1	2	1	9	V
Arabis scabra	—	—	1/1	0	2	4	1	0	2	2	9	V
Atriplex longipes	—	—	2/3	1	3	3	0	2	1	2	9	R
Buglossoides purpurocaerulea	—	—	11/22	1	13	1	2	1	2	2	9	R
Calamagrostis scotica	—	—	2/4	1	2	4	0	2	1	1	9	V
Centaurium scilloides	—	—	2/3	1	6	2	1	1	2	2	9	V
Chenopodium vulvaria	—	2/3	15/95	2	15	1	0	2	2	2	9	V
Corrigiola litoralis	—	—	1/2	1	1	4	0	0	2	2	9	V
Crassula aquatica	—	—	1/2	1	1	4	0	2	1	1	9	V
Crepis foetida	—	—	2/19	2	4	3	0	1	1	2	9	V
Echium plantagineum	—	1/1	4/6	1	6	2	1	2	1	2	9	V
Equisetum ramosissimum	—	—	1/1	0	1	4	0	2	1	2	9	E
Euphorbia peplis	0/1	2/5	1/22	2	1	4	0	2	0	1	9	E
Filago lutescens	—	—	10/67	2	15	1	0	2	2	2	9	V
F. pyramidata	—	0/1	11/96	2	11	1	0	2	2	2	9	V
Fritillaria meleagris	—	—	15/116	2	20	0	2	1	2	2	9	V
Galium fleurotii	—	—	1/1	0	1	4	1	1	2	1	9	R
Gastridium ventricosum	—	2/2	6/125	2	6	2	0	1	2	2	9	V
Iris versicolor	—	—	6/10	1	8	2	2	1	1	2	9	R
Isatis tinctoria	—	—	2/2	0	2	4	1	1	2	1	9	V
Juncus subulatus	—	—	1/1	0	1	4	0	1	2	2	9	R
Leersia oryzoides	—	—	7/21	2	8	2	0	2	2	1	9	V
Limonium paradoxum	1/1	—	1/1	0	1	4	1	2	1	1	9	E
L. transwallianum	2/3	—	2/3	1	2	4	1	1	1	1	9	R
Liparis loeselii	—	—	8/29	2	8	2	1	1	1	2	9	V
Lobelia urens	—	—	8/12	1	10	1	2	1	2	2	9	V
Lotus angustissimus	—	5/5	13/46	2	14	1	1	1	2	2	9	R

SPECIES	1 Ireland (H)	2 Channel Isles (S)	3 Great Britain (GB)	I	4 Total	I	5 Attractiveness I	6 Conservation I	7 Remoteness I	8 Accessibility I	9 TN	10 IUCN Category
	1	2	3		4		5	6	7	8	9	10

TN = 9 (53 species) — *continued*

SPECIES	1	2	3	I	Total	I	5	6	7	8	9	10
Minuartia stricta	—	—	1/1	0	1	4	1	1	1	2	9	V
Muscari atlanticum	—	—	10/17	1	17	0	2	2	2	2	9	V
Narcissus obvallaris	—	—	7/9	0	7	2	2	1	2	2	9	R
Neotinea maculata	19/32	—	1/1	0	1	4	2	2	0	1	9	R
Oenothera stricta	—	4/4	10/29	2	12	1	2	1	1	2	9	R
Ophrys fuciflora	—	—	4/6	1	10	1	2	1	2	2	9	R
Paeonia mascula	—	—	1/2	1	1	4	2	0	0	2	9	V
Phyllodoce caerulea	—	—	3/3	0	4	3	2	1	1	2	9	V
Polygonatum verticillatum	—	—	4/10	1	5	3	2	1	1	1	9	V
Polygonum maritimum	1/1	1/3	2/11	2	2	4	0	2	0	1	9	E
Potentilla rupestris	—	—	3/3	0	3	3	2	2	1	1	9	V
Ranunculus ophioglossifolius	—	0/1	2/4	1	2	4	0	1	1	2	9	E
Rhinanthus serotinus	—	—	5/68	2	10	1	1	2	1	2	9	V
Selinum carvifola	—	—	2/5	1	2	4	0	1	2	1	9	V
Senecio cambrensis	—	—	5/5	0	6	2	1	2	2	2	9	R
Taraxacum acutum	—	—	2/2	0	2	4	0	2	1	2	9	V
T. austrinum	1/1	1/2	1/2	1	2	4	0	1	1	2	9	V
T. glaucinum	—	—	2/4	1	2	4	0	1	1	2	9	V
Tetragonolobus maritimus	—	—	9/9	0	9	2	1	2	2	2	9	R
Trichomanes speciosum	22/47	—	8/15	1	8	2	2	2	1	1	9	V
Trifolium bocconei	—	1/1	2/3	1	5	3	1	1	1	2	9	R
Valerianella rimosa	9/41	—	9/96	2	10	1	0	2	2	2	9	V
Veronica verna	—	—	1/8	2	8	2	0	1	2	2	9	E
Woodsia ilvensis	—	—	4/12	2	4	3	2	1	0	1	9	V

SPECIES	1	2	3		4		5	6	7	8	9	10
	Ireland (H)	Channel Isles (S)	Great Britain (GB)	**I**	Total **I**		Attractiveness **I**	Conservation **I**	Remoteness **I**	Accessibility **I**	**TN**	IUCN Category

Rate of decline (cols 1–3); Local'es (col 4)

TN = 8 (38 species)

SPECIES	1	2	3		4		5	6	7	8	9	10
Alisma gramineum	—	—	4/4	0	4	3	1	1	2	1	8	R
Astragalus alpinus	—	—	4/4	0	4	3	2	1	1	1	8	R
Atriplex praecox	—	—	3/3	0	3	3	0	2	1	2	8	R
Bromus madritensis	1/2	4/4	1/4	2	3	3	0	1	1	1	8	R
Calamintha sylvatica	—	—	1/1	0	1	4	2	0	1	1	8	V
Carex buxbaumii	0/1	—	2/2	0	2	4	0	2	1	1	8	V
Centaurea calcitrapa	—	0/2	14/109	2	16	0	1	1	2	2	8	R
Crocus purpureus	1/1	—	8/11	0	11	1	2	1	2	2	8	R
Dryopteris cristata	—	—	9/26	1	21	0	2	2	1	2	8	V
Epipactis dunensis	—	—	8/8	0	11	1	1	2	2	2	8	R
Erica ciliaris	1/1	—	9/17	1	44	0	2	1	2	2	8	R
E. vagans	1/2	—	6/8	0	10	1	2	1	2	2	8	R
Eriophorum gracile	6/6	—	5/13	1	6	2	0	2	2	1	8	V
Euphorbia hyberna	86/103	—	2/2	0	3	3	1	1	1	2	8	R
Gentianella uliginosa	—	—	5/5	0	8	2	2	0	2	2	8	V
Helianthemum apenninum	—	—	4/4	0	8	2	2	0	2	2	8	R
Herniaria ciliolata	—	3/3	2/4	1	9	2	0	1	2	2	8	R
H. glabra	—	—	8/15	1	12	1	0	2	2	2	8	V
Juncus mutabilis	—	—	2/3	1	4	3	0	1	1	2	8	R
Leucojum aestivum	1/4	—	14/29	1	37	0	2	2	2	1	8	R
Lloydia serotina	—	—	2/2	0	5	3	2	1	2	0	8	V
Luzula pallescens	1/1	—	2/2	0	2	4	0	0	2	2	8	R
Maianthemum bifolium	—	—	5/12	1	6	2	2	1	1	1	8	V
Najas marina	—	—	2/3	1	4	3	0	2	2	0	8	V
Ononis reclinata	—	2/3	5/7	1	5	3	1	1	1	2	8	V
Orobanche maritima	—	4/5	13/35	1	12	1	1	1	2	2	8	V
O. purpurea	—	3/4	8/19	1	11	1	1	2	2	1	8	V
Phyteuma spicatum	—	—	6/6	0	10	1	2	1	2	2	8	R
Polycarpon tetraphyllum	—	5/5	3/13	2	8	2	0	1	1	2	8	R
Rumex rupestris	—	3/4	11/27	1	12	1	1	1	2	2	8	V
Saxifraga cernua	—	—	3/5	1	4	3	2	1	0	1	8	V
Seseli libanotis	—	—	4/6	1	7	2	0	1	2	2	8	R
Teucrium botrys	—	—	8/11	0	8	2	1	1	2	2	8	R
Trifolium molinerii	—	1/1	2/3	1	7	2	1	1	1	2	8	R
T. strictum	—	1/2	2/4	1	6	2	1	1	1	2	8	R
Valerianella eriocarpa	—	1/2	10/31	2	13	1	0	2	1	2	8	V
Veronica praecox	—	—	3/3	0	8	2	0	2	2	2	8	E
Viola kitaibeliana	—	3/3	1/1	0	5	3	1	1	1	2	8	R

	1	2	3		4		5	6	7	8	9	10
		Rate of Decline			Local'es							
SPECIES	Ireland (H)	Channel Isles (S)	Great Britain (GB)	**I**	Total	**I**	Attractiveness **I**	Conservation **I**	Remoteness **I**	Accessibility **I**	**TN**	IUCN Category

TN = 7 (30 species)

	1	2	3		4		5	6	7	8	9	10
Alchemilla glaucescens	1/1	—	9/15	1	13	1	1	1	1	2	7	R
Arabis alpina	—	—	1/1	0	2	4	2	1	0	0	7	R
Bromus tectorum	—	—	5/5	0	10	1	0	2	2	2	7	R
Buxus sempervirens	—	—	7/8	0	28	0	2	1	2	2	7	R
Cerastium arcticum subsp. edmondstonii	—	—	2/2	0	4	3	1	1	0	2	7	R
C. brachypetalum	—	—	1/1	0	4	3	0	2	2	0	7	R
Euphorbia serrulata	—	—	5/5	0	10	1	1	1	2	2	7	R
Euphrasia cambrica	—	—	2/3	1	6	2	1	1	1	1	7	R
E. rotundifolia	—	—	9/14	1	9	2	1	2	0	1	7	R
E. vigursii	—	—	13/25	1	19	0	1	2	1	2	7	R
Gentiana nivalis	—	—	4/4	0	9	2	2	1	1	1	7	V
Gladiolus illyricus	—	—	7/9	0	40	0	2	1	2	2	7	R
Homogyne alpina	—	—	2/2	0	2	4	1	1	0	1	7	R
Juncus capitatus	—	5/5	5/10	1	5	3	0	1	1	1	7	R
J. filiformis	—	—	9/15	1	12	1	0	1	2	2	7	R
J. nodulosus	—	—	2/2	0	2	4	0	1	1	1	7	R
Lavatera cretica	—	2/2	1/3	1	13	1	1	1	1	2	7	R
Ludwigia palustris	—	0/1	5/9	1	13	1	0	1	2	2	7	R
Lychnis alpina	—	—	3/4	0	4	3	2	1	0	1	7	R
Mibora minima	—	3/4	4/4	0	5	3	0	1	1	2	7	R
Moneses uniflora	—	—	13/24	1	17	0	2	1	1	2	7	R
Ophioglossum lusitanicum	—	1/1	1/1	0	1	4	0	1	1	1	7	R
Physospermum cornubiense	—	—	7/11	1	∞	0	0	2	2	2	7	R
Potamogeton nodosus	—	—	7/11	1	10	1	0	2	2	1	7	R
Rorippa austriaca	0/1	—	13/15	0	13	1	0	2	2	2	7	R
Sorbus eminens	—	—	2/3	1	3	3	1	0	2	0	7	R
S. wilmottiana	—	—	1/1	0	2	4	1	0	1	1	7	R
Taraxacum ziphoideum	—	—	3/3	0	3	3	0	2	1	1	7	R
Thlaspi perfoliatum	—	—	7/14	1	11	1	0	1	2	2	7	R
Woodsia alpina	—	—	11/17	1	15	1	2	1	1	1	7	R

SPECIES	Rate of decline				Local'es		Attractiveness	Conservation	Remoteness	Accessibility	TN	IUCN Category
	Ireland (H)	Channel Isles (S)	Great Britain (GB)	I	Total	I	I	I	I	I	TN	
	1	2	3		4		5	6	7	8	9	10

TN = 6 (49 species)

SPECIES	1	2	3		4		5	6	7	8	9	10
Alchemilla acutiloba	—	—	11/11	0	∞	0	1	2	1	2	6	R
Arenaria norvegica subsp. norvegica	1/1	—	8/8	0	12	1	1	1	1	2	6	R
Carex chordorrhiza	—	—	1/1	0	2	4	0	1	0	1	6	R
C. filiformis	—	—	9/13	0	10	2	0	1	2	1	6	R
C. flava	—	—	1/1	0	1	4	0	0	1	1	6	R
C. recta	—	—	3/4	0	6	2	0	2	1	1	6	R
Cirsium tuberosum	—	—	11/13	0	20	0	1	1	2	2	6	R
Cochlearia micacea	—	—	3/3	0	3	3	0	1	1	1	6	R
Cynodon dactylon	—	1/1	10/12	0	11	1	0	2	1	2	6	R
Euphrasia campbelliae	—	—	7/7	0	9	2	1	2	0	1	6	R
E. heslop-harrisonii	—	—	10/11	0	12	1	1	2	0	2	6	R
E. rhumica	—	—	1/1	0	2	4	1	0	0	1	6	R
Fumaria occidentalis	—	—	11/12	0	23	0	1	1	2	2	6	R
Galium debile	—	0/2	6/6	0	17	0	0	2	2	2	6	R
Gentiana verna	20/26	—	5/6	0	∞	0	2	1	1	2	6	R
Gnaphalium norvegicum	—	—	9/14	1	9	2	1	0	1	1	6	R
Helianthemum canum subsp. levigatum	—	—	1/1	0	1	4	1	0	0	1	6	R
Hypochoeris maculata	—	1/1	9/16	1	13	1	1	1	1	1	6	R
Isoetes histrix	—	2/2	2/2	0	7	2	0	1	1	2	6	R
Ledum groenlandicum	—	—	9/10	0	12	1	2	1	1	1	6	R
Limonium bellidifolium	—	—	6/10	1	23	0	1	0	2	2	6	R
Lynchis viscaria	—	—	12/19	1	16	0	2	1	1	1	6	R
Matthiola incana	—	—	3/4	0	7	2	2	1	1	0	6	R
Minuartia rubella	—	—	5/6	0	6	2	0	2	1	1	6	R
Myosotis alpestris	—	—	6/7	0	11	1	2	1	1	1	6	R

SPECIES	1 Ireland (H)	2 Channel Isles (S)	3 Great Britain (GB)	I	4 Total	I	5 Attractiveness I	6 Conservation I	7 Remoteness I	8 Accessibility I	9 TN	10 IUCN Category
			Rate of Decline		Local'es							

TN = 6 (49 species) — *continued*

SPECIES	1	2	3		4		5	6	7	8	9	10
Ornithopus pinnatus	—	3/5	1/2	1	17	0	1	1	1	2	6	R
Oxytropis campestris	—	—	3/3	0	6	2	2	1	1	0	6	R
O.　　halleri	—	—	11/17	1	21	0	2	1	1	1	6	R
Peucedanum officinale	—	—	5/6	0	15	1	1	1	2	1	6	R
Phleum phleoides	—	—	10/22	1	30	0	0	1	2	2	6	R
Polygala amara	—	—	6/6	0	13	1	1	1	1	2	6	R
P.　　austriaca	—	—	7/9	0	13	1	1	1	1	2	6	R
Potentilla fruticosa	5/8	—	6/7	0	12	1	2	1	1	1	6	R
Rumex aquaticus	—	—	3/3	0	9	2	0	1	2	1	6	R
Sagina intermedia	—	—	4/6	1	4	3	0	0	0	2	6	R
Saxifraga cespitosa	—	—	9/9	0	9	2	2	1	0	1	6	R
Scirpus holoschoenus	—	—	2/2	0	3	3	0	1	1	1	6	R
Scrophularia scorodonia	—	3/3	10/25	1	17	0	0	1	2	2	6	R
Silene otites	—	—	12/19	1	21	0	0	1	2	2	6	R
Sorbus arranensis	—	—	1/1	0	2	4	1	0	1	0	6	R
S.　　bristoliensis	—	—	1/1	0	3	3	1	0	1	1	6	R
S.　　lancastriensis	—	—	3/4	0	6	2	1	2	1	0	6	R
S.　　leptophylla	—	—	2/2	0	3	3	1	0	1	1	6	R
S.　　leyana	—	—	2/2	0	2	4	1	0	1	0	6	R
S.　　pseudofennica	—	—	1/1	0	1	4	1	0	1	0	6	R
Taraxacum cymbifolium	—	—	1/1	0	1	4	0	0	1	1	6	R
T.　　hygrophilum	—	—	1/1	0	6	2	0	1	1	2	6	R
Thymus serpyllum	—	—	5/6	0	18	0	1	1	2	2	6	R
Viola rupestris	—	—	6/6	0	11	1	1	1	1	2	6	R

TN = 5 (29 species)

SPECIES	Ireland (H)	Channel Isles (S)	Great Britain (GB)	**I**	Total	**I**	Attractiveness	Conservation	Remoteness	Accessibility	**TN**	IUCN Category
	1	2	3		4		5	6	7	8	9	10
Alchemilla monticola	—	—	9/9	0	∞	0	1	1	1	2	5	R
Allium babingtonii	4/18	—	12/16	0	26	0	1	1	1	2	5	R
Artemisia norvegica	—	—	3/3	0	3	3	0	1	0	1	5	R
Aster linosyris	—	—	6/7	0	6	2	1	0	1	1	5	R
Bunium bulbocastanum	—	—	11/15	0	∞	0	0	1	2	2	5	R
Carex norvegica	—	—	3/4	0	4	3	0	2	0	0	5	R
Cicerbita alpina	—	—	4/4	0	4	3	2	0	0	0	5	R
Draba aizoides	—	—	3/3	0	12	1	2	0	1	1	5	R
Elatine hydropiper	11/14	—	6/15	1	8	2	0	1	1	0	5	R
Erigeron borealis	—	—	6/9	1	10	1	1	0	1	1	5	R
Euphrasia eurycarpa	—	—	1/1	0	3	3	1	0	0	1	5	R
E. rivularis	—	—	9/11	0	10	1	1	1	1	1	5	R
Festuca caesia	—	2/2	4/4	0	10	2	0	1	1	1	5	R
Genista pilosa	—	—	11/19	1	42	0	1	1	1	1	5	R
Juncus dudleyi	—	—	2/2	0	2	4	0	1	0	0	5	R
Koeleria vallesiana	—	—	4/4	0	12	1	0	1	1	2	5	R
Limosella australis	—	—	3/3	0	8	2	0	1	1	1	5	R
Poa infirma	—	5/5	5/5	0	12	1	0	1	1	2	5	R
Polemonium caeruleum	—	—	15/18	0	27	0	2	1	2	0	5	R
Potamogeton epihydrus	—	—	2/2	0	3	3	0	1	0	1	5	R
Rhynchosinapis wrightii	—	—	1/1	0	2	4	0	1	0	0	5	R
Sagina normaniana	—	—	6/18	2	7	2	0	0	0	1	5	R
Saxifraga hirculus	2/11	—	10/21	1	21	0	2	1	0	1	5	R
S. rivularis	—	—	12/16	1	17	0	2	1	1	0	5	R
Sorbus minima	—	—	1/1	0	3	3	1	0	1	0	5	R
Spiranthes romanzoffiana	11/19	—	10/12	0	12	1	2	1	0	1	5	R
Taraxacum pseudonordstedtii	—	—	1/1	0	6	2	0	0	1	2	5	R
Trinia glauca	—	—	5/5	0	10	1	1	0	1	2	5	R
Veronica fruticans	—	—	14/20	0	20	0	2	1	1	1	5	R

Columns headings: 1–2 Ireland (H) / Channel Isles (S); 1 2 3 **Rate of decline**; 4 **Local'es**; Total

SPECIES	1 Ireland (H)	2 Channel Isles (S)	3 Great Britain (GB)	I	4 Total	I	5 Attractiveness	6 Conservation	7 Remoteness	8 Accessibility	9 TN	10 IUCN Category
			Rate of Decline		Local'es							

TN = 4 (12 species)

SPECIES	1	2	3	I	4	I	5	6	7	8	9	10
Bartsia alpina	—	—	12/15	0	19	0	1	1	1	1	4	R
Carex atrofusca	—	—	4/5	0	4	3	0	1	0	0	4	R
C. lachenalii	—	—	6/7	0	7	2	0	1	1	0	4	R
C. microglochin	—	—	1/1	0	1	4	0	0	0	0	4	R
C. ornithopoda	—	—	10/11	0	15	1	0	1	1	1	4	R
C. rariflora	—	—	7/12	1	7	2	0	1	0	0	4	R
Eleocharis austriaca	—	—	10/10	0	17	0	0	2	1	1	4	R
E. parvula	3/3	—	7/10	0	9	2	0	1	1	0	4	R
Eriocaulon aquaticum	37/52	—	7/9	0	10	1	1	1	0	1	4	R
Potamogeton rutilus	—	—	9/9	0	12	1	0	2	0	1	4	R
Salix lanata	—	—	10/11	0	14	1	1	1	0	1	4	R
Sorbus porrigentiformis	—	—	14/19	0	20	0	1	1	2	0	4	R

TN = 3 (6 species)

SPECIES	1	2	3	I	4	I	5	6	7	8	9	10
Agropyron donianum	—	—	11/11	0	16	0	0	1	1	1	3	R
Hierochloe odorata	1/1	—	11/12	0	17	0	0	1	1	1	3	R
Kobresia simpliciuscula	—	—	10/14	0	11	1	0	1	1	0	3	R
Koenigia islandica	—	—	5/5	0	6	2	0	1	0	0	3	R
Poa flexuosa	—	—	7/7	0	7	2	0	1	0	0	3	R
Sorbus anglica	1/1	—	12/12	0	16	0	1	1	1	0	3	R

TN = 2 (1 species)

SPECIES	1	2	3	I	4	I	5	6	7	8	9	10
Najas flexilis	8/10	—	12/15	0	12	1	0	1	0	0	2	R

TABLE 2

Habitats of species that are nationally rare or threatened in Great Britain

	Extinct	Endangered	Vulnerable	Rare	Total
Montane	2(3)	0(0)	10(17)	48(80)	60(19)
Woodland, scrub and hedge	1(3)	5(14)	14(40)	15(43)	35(11)
Arable	1(4)	14(62)	7(30)	1(4)	23(7)
Man-made open habitats, road-sides, quarries, etc	4(16)	5(19)	7(27)	10(38)	26(8)
Lowland pasture, open grassland, natural open habitats	1(1)	7(10)	28(40)	35(49)	71(22)
Heath	0(0)	3(18)	3(18)	11(64)	17(5)
Wetlands	6(13)	7(15)	16(35)	17(37)	46(14)
Aquatic	1(8)	0(0)	2(17)	9(75)	12(4)
Maritime	3(10)	5(16)	8(26)	15(48)	31(10)
Total	19(6)	46(14)	95(30)	161(50)	321

INDEX

Bold type indicates Threat Number and IUCN Category

95